Introductory Physical Geology
Laboratory Manual for Distance Learning

GREG P. GARDINER

SUSAN WILCOX

Kendall Hunt
publishing company

D1300693

Kendall Hunt Book Team:
publishing company

Mark C. Falb, Chairman and Chief Executive Officer
Chad M. Chandlee, President and Chief Operating Officer
David L. Tart, Vice President, Higher Education
Paul B. Carty, Director of Publishing Partnerships
Georgia Botsford, Editorial Manager
Lynne Rogers, Senior Editor
Timothy J. Beitzel, Vice President, Operations
Christine E. O'Brien, Assistant Vice President, Production Services
Mary Melloy, Senior Production Editor
Renae Horstman, Senior Permissions Coordinator

Coast Book Team:

Andrew C. Jones, Chancellor, Coast Community College District
Loretta P. Adrian, President, Coastline Community College
Dan C. Jones, Executive Dean, Office of Instructional Systems Development
Lynn M. Dahnke, Director, Marketing & Publisher Partnerships
Robert D. Nash, Director, Instructional Design & Faculty Support
Sylvia E. Amito'elau, Educational Media Designer
Judy Garvey, Director, E-Media & Publishing
Wendy Sacket, E-Media & Publishing Project Coordinator
Marie Hulett, Production Coordinator
Thien Vu, E-Media & Publishing Assistant

Figures 5.1, 5.4, 6.2, and Appendix V adapted from *Geology Laboratory for Distance Learning* by James L. Ruhle. Copyright © 2000 by the Coast Community College District. Reprinted by permission of Kendall Hunt Publishing Company.

Figures 2.1, 2.2, 2.4, 2.8, 2.9, 2.13, 3.8, 3.9, 3.13, 4.9, 5.6, 6.5, 7.2, 7.4, 8.5, 9.5, 10.8, 10.9A, 11.2, 12.3, 12.7A & B, and 12.10 from *Planet Earth* by John J. Renton. Copyright © 2002 by John J. Renton. Reprinted by permission of Kendall Hunt Publishing Company.

Figures 2.5, 2.10, 2.11, 2.14, 2.15, 3.7, 4.8, 5.2, 5.3, 8.2, 8.3, 9.4, 9.10, 10.5A, 11.3, 11.4, 11.5A, 11.6, 11.7, 11.8, 11.9, 11.10, 11.12, 12.2, 12.8, 12.11, 13.1, 13.2A, 13.4, 13.5, 13.15, 13.16B, and 13.19 A & B from *Physical Geology Across the American Landscape*, 3d edition, by John J. Renton. Copyright © 2011 by the Coast Community College District and Kendall Hunt Publishing Company.

Cover design by Don Vierstra; cover images © iStock, Inc., and Shutterstock, Inc. All Shutterstock images used under license from Shutterstock, Inc.

ISBN-13: 978-0-7575-6320-1

Coast Learning Systems
Coastline Community College
11460 Warner Avenue
Fountain Valley, CA 92708
fax: (714) 241-6286   e-mail: CoastLearning@coastline.edu   website: www.CoastLearning.org

Printed in the United States of America
10 9 8 7 6 5 4 3

Contents

Acknowledgments

Several of the individuals responsible for the creation of this course are listed on the copyright page of this book. In addition to these people, appreciation is expressed for the contributions of the following individuals:

Susan Wilcox, M.A., considers it a personal mission to get liberal arts students excited about science. Her own love of the sciences is such that she once strongly considered changing her major to Physics and Astronomy. Instead, she graduated magna cum laude from the University of Wisconsin-Milwaukee with a Bachelor of Arts degree in Mass Communications and Spanish, followed by a Master of Arts degree in Communications from the University of Hawaii.

Over the course of the next decade, Susan worked in broadcasting, print, and corporate communications. She became adept at designing messages that communicated complex, high-tech subjects to "low-tech audiences." She also developed and taught courses at the University of Hawaii and at Leeward Community College in Honolulu. Her leisure time was spent delving into the physical, natural, and social sciences.

Susan moved to California in 1988, where she continued her media work. A decade later, she and Coast Learning Systems found each other; it was a perfect fit. Since then, Susan has been a writer, producer, and project manager for algebra, psychology, child development, anthropology, biology, statistics, chemistry, and geology courses. She especially enjoys translating scientific subjects into concepts that can be understood and appreciated by the nonscientific mind, emphasizing that science is behind everything that happens in our world.

Greg P. Gardiner, M.S., has more than 12 years experience as an educator teaching geology and earth science. He has a teaching credential in biological sciences, a supplemental credential in geological sciences, and holds a Master's degree in Environmental Science from California State University, Fullerton. Greg has presented a paper at the California Academy of Sciences. He has a passion for geologic sciences and has worked diligently to coordinate the instructional content of these lab lessons by linking the exercises to real-life activities in order to help students understand the concepts and processes of this fascinating subject. Greg has also conducted field study programs at Yosemite National Park and Catalina Island, California.

Sylvia E. Amito'elau. M.S., has overseen the instructional design of this lab manual, as well its accompanying textbook and online course, from concept to completion. She is an instructional designer for Coast Learning Systems, a division of Coastline Community College in Fountain Valley, California. She has assisted in design and development on several educational projects, including online courses in accounting, Arabic, chemistry, Chinese, education, math, and student success for more than 8 years. At Coastline Community College, Sylvia is responsible for providing instructional design, training, and support for all faculty, particularly in areas related to distance learning. As a member of the Senate Academic Standards Committee, she participated in the development of the Coastline Academic Quality Rubric. She is also a part-time faculty member teaching computer application courses and has experience teaching courses in various delivery modalities such as classroom, hybrid, and online. In addition, Sylvia has worked on the California Virtual Campus project, training and assisting Southern California community college faculty in the design, development, and delivery of online instruction. Sylvia holds a Master of Science degree in Instructional Technology and a Bachelor of Arts degree in Mathematics.

We would like to express our thanks to the members of the Academic Advisory Team whose names appear below. In particular, we would like to express special thanks to Debbie Secord, M.S., and Jay Yett, M.S., for their dedication to reviewing the content of this manual.

Special thanks are also owed to the graphic design contributions made by Bob Dixon, Marie Hulett, Don Vierstra, and Mark Worden.

Academic Advisory Team

Robert Altamura, Ph.D., Florida Community College at Jacksonville Open Campus, Urban Resources Center

Edward (Erik) Bender, M.S., Orange Coast College

Theodore Erski, M.A., McHenry County College

Roberto Falero, M.S., DPRA, Inc.

Gail Gibson, Ph.D., Florida Community College at Jacksonville—Kent Campus

Jonathan Kuespert, M.S., M.B.A., BreitBurn Energy Management Company

Michael Leach, M.S., M.A., New Mexico State University

James McClinton, M.S., Eastern New Mexico University—Roswell

Joseph Mraz, M.S., Santa Fe Community College

Jay P. Muza, Ph.D., Broward College

Douglas Neves, Ph.D., Cypress College

Kathy Ann Randall, M.S., Lincoln County Campus of the Flathead Valley Community College

Kelly Ruppert, M.S., California State University, Fullerton, and Coastline Community College

Richard Schultz, Ph.D., C.P.G., Elmhurst College

Debbie Secord, M.S., Coastline Community College

William H. Walker, Ph.D., Thomas Edison State College

Curtis Williams, M.S., California State University, Fullerton

Jan Yett, M.S., Orange Coast College

Preface: How to Take This Lab Course

To the Student

Welcome to the *Introductory Physical Geology Laboratory Manual for Distance Learning*. The first lab lesson in this manual deals with how to read and use topographic maps. The remaining lessons follow a sequence that progresses through the basics of plate tectonics, seismology, minerals and rocks, and geologic time and concludes with such overarching topics as Earth's major geologic features and economic geology resources.

Learning Outcomes

The designers, academic advisors, and producers of this lab manual have specified the following learning outcomes for students using the *Introductory Physical Geology Laboratory Manual for Distance Learning*. After successfully completing the lab exercises, you should be able to:

1. Effectively apply the concepts, principles, and theories of geology to make accurate observations and to identify and distinguish among samples/structures/landscapes.
2. Gather and analyze data, formulate and test hypotheses, solve problems, and come to supportable conclusions given various scenarios and research topics.

Features

This manual and laboratory kit are part of an intensive laboratory course that explores the basic concepts and principles of physical geology. Each lesson includes specific learning objectives that students should use to prepare for the lab. The lab manual includes exercises and procedures that illuminate the central principles of physical geology. Each lab lesson includes questions designed to help you analyze, review, and apply your knowledge of the material covered in the lab course. Reading this lab manual, watching the video clips and completing the activities in the online component, and completing the lab exercises will provide you with information that you would receive in the classroom if you were taking this lab course on campus.

The laboratory kit contains most of the materials and mineral samples necessary to conduct the lab exercises contained in each lesson. Each lesson in the lab manual contains the following elements:

➤ Overview

This section introduces the topics covered in the lab lesson, explains why they are important, and makes connections to previous lesson concepts that you'll need to remember.

➤ Learning Objectives

These objectives outline the significant goals to be achieved after completing each lesson. (Note: Instructors often design test questions based on learning objectives, so use them to help focus your study.)

➤ Materials

This section provides a list of materials that will be needed to complete the lab exercises. Some items will be provided in the accompanying lab kit, and others may need to be purchased or borrowed if they are not readily available in your home.

➤ Illustrations

These drawings and photographs have been included to amplify your understanding of specific concepts or to illustrate particular steps and procedures within the course of various lab exercises.

➤ Lab Exercises

The Lab Exercises section focuses on structured investigations and experiments with materials included in the lab kit as well as case-based learning scenarios.

➤ Online Activities

The Online Activities section involves using the Internet to access the course website, where you may watch videos, view images, upload documents, and complete quizzes according to your instructor's directions.

➤ Quiz

This section includes a variety of questions designed to verify your comprehension of the lab lessons and will help you make connections to and apply the principles covered within the course.

How to Take This Distance Learning Lab

If this is your first experience with distance learning, welcome. Distance learning courses are designed for busy people whose situations or schedules do not permit them to take a traditional on-campus course.

This lab manual has been designed to be used as a tool to help reinforce topics and concepts on which you will later be tested. To complete this lab course successfully, you will need to complete exercises that:

- provide you with information that you can apply to your everyday experiences.
- provide visual reinforcement to help you understand and appreciate the complexity of the various physical geologic processes that occur above and beneath the surface of the earth as you know it.
- provide you with the opportunity to practice what you have learned.

- help make the study of physical geology more organized, systematic, and enjoyable. Since you are required to assimilate a large amount of information in a short period of time, a lot of your dedicated time is required. You should be prepared to set aside time when you can tackle and complete an entire lab exercise so that you can master the concepts involved and be prepared for assessment.

Even though you do not have scheduled classes to attend each week on campus, please keep in mind that this is a college-level course. It will require the same amount of work as a traditional, classroom version of this lab course and at the same level of difficulty. As a distance learner, however, it will be up to you alone to keep up with your deadlines. It's important that you schedule enough time to read, study, review, and reflect. Also, take some time immediately after completing a lab lesson to reflect on what you have just learned. This is an excellent time to discuss the lesson with a friend or family member. Your active thinking and involvement will promote your success.

TOPOGRAPHIC MAPS

Lesson 1

AT A GLANCE

Purpose

Learning Objectives

Materials Needed

Overview

    Topographic Maps

    Compass Bearings

    Public Land Survey System

    Scale

    Scale Conversion

    Map Symbols

    Contour Lines

Online Activities

Lab Exercises

    Lab Exercise #1: Contour Exercises

    Lab Exercise #2: Topographic Profiles

Quiz

    Multiple Choice

    Short Answer

Purpose

The activities in this lesson will lay the foundation for the use of topographic maps. Knowing how to read and interpret a topographic map is essential for understanding geological features in any area. Hikers, campers, geologists, and engineers use the topographic map as a two-dimensional tool to allow them to interpret three-dimensional land surfaces for their scientific studies or recreational use.

Learning Objectives

After completing this laboratory lesson, you will be able to:

- Understand how specific elevation points relate to the overall topography of a broad area.
- Understand how different landforms are depicted on maps.
- Understand how to interpret map symbols and scales.
- Construct a topographic map by drawing contour lines based on points of elevation.
- Construct a topographic profile.
- Read and interpret topographic maps.

Materials Needed

The activities will be performed using the following materials provided in the lab kit, as well as materials readily available in your home. If you do not already have them in your home, you may need to purchase some of the materials. Be sure you have all listed materials before starting the activity.

- ❑ A USGS topographic map of Yosemite Valley, California (included in your laboratory kit)
- ❑ USGS topographic map symbols pamphlet (included in your laboratory kit)
- ❑ Protractor or compass (included in your laboratory kit)
- ❑ Ruler (included in your laboratory kit)
- ❑ Pencil
- ❑ Eraser
- ❑ Calculator

Overview

A topographic map is a flat, two-dimensional representation of a three-dimensional land surface. Unlike the more familiar atlas that illustrates the locations of towns and roads, a topographic map illustrates the peaks, hills, slopes, and depressions of a landscape in a way that an informed user can interpret, using individual contour lines to depict elevation information. Elevation is the same at all points along a single contour line; a series of contour lines indicates a change in elevation—a slope—as well as steepness. Topographic maps also include information about streams, bodies of water, towns, buildings, campgrounds, boundaries, and other natural and man-made features.

In this lesson, you'll learn how topographic maps are made and how to read them, a task that falls to any scientist working in the field as well as to many hikers and adventurers. A topographic map can tell you much about an area. Knowing how to read one can help you not only get where you want to go, but can also help you determine your location if you are lost. Remember, cell phones with GPS don't work everywhere!

Take the topographic map from your lab kit and unfold it on the table, so that as you read through this overview, you can refer to an actual map.

Topographic Maps

Most topographic maps of areas within the United States are published by the United States Geological Survey (USGS), and most of these depict rectangular sections of the earth's surface known as quadrangles. The exact location of a quadrangle is expressed in terms of its latitude and longitude.

Latitude and longitude are imaginary lines that divide Earth's surface in a way that allows any location to be pinpointed by stating its coordinates. The coordinates are expressed in terms of degrees, minutes, and seconds away from a baseline, or "starting line."

Figure 1.1 illustrates latitude and longitude. Latitude is the value that describes how far north or south a location is from the baseline of the equator. As you may recall, the equator is an imaginary line encircling Earth perpendicular to its axis of rotation at a location that is equidistant from the poles; it divides the earth into two equal hemispheres, the northern and southern hemispheres. Lines of latitude encircle Earth parallel to the equator, so they are often called parallels. Each parallel has a value that describes its distance north or south of the equator, with the equator having a value of 0° latitude and the poles each having the values of 90° latitude. Their actual values are: North Pole, 90° N; South Pole, 90° W.

Longitude is a value that describes how far east or west a location is from a north-south baseline, the prime meridian that passes through Greenwich England. Lines of longitude, also called meridians, do not encircle Earth but extend from pole to pole, dividing Earth like the sections of an orange. The prime meridian has a value of 0° longitude, with numbers increasing toward the east (east longitude) and toward the west (west longitude) until they reach the opposite side of Earth, which is the International Date Line, a meridian with the value of 180° longitude. (Note that 180° plus 180° equals 360°, the number of degrees in a circle.)

As seen in **Figure 1.1**, lines of longitude are not like latitude lines but converge at the North Pole and the South Pole. Since North America is north of the equator and west of the prime meridian, all latitudes in the continental United States are north latitudes and all longitudes are west longitudes.

To pinpoint a location, degrees of latitude and longitude are further divided into minutes and seconds:

- 1 degree (°) = 60 minutes (')
- 1 minute (') = 60 seconds (")

Latitude and longitude coordinates are expressed as follows: Degrees, minutes, seconds plus direction from either the equator or prime meridian. For instance, the coordinates of South Dakota's Mount Rushmore are:

$$43° \ 52' \ 49" \ N; \ 103° \ 27' \ 33" \ W$$

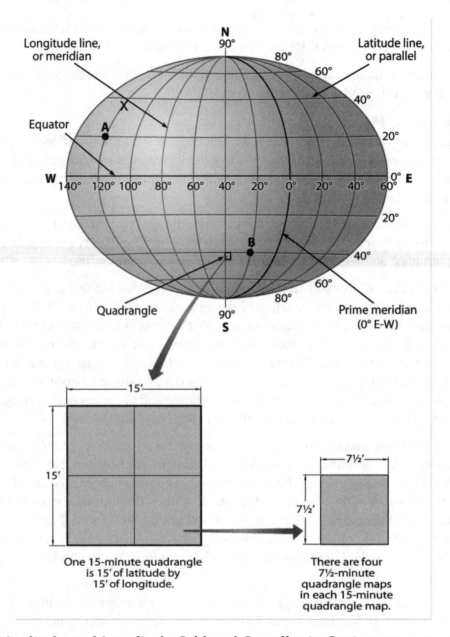

Figure 1.1 Latitude and Longitude Grid and Coordinate System. Illustration by Don Vierstra.

Practice a little with **Figure 1.1**. Answers appear in the student answer key at the end of this lesson.

What is the latitude and longitude value of Point A?

Latitude: _____

Longitude: _____

What about Point B?

Latitude: _____

Longitude: _____

Lesson 1/ Topographic Maps

Now, look at your Yosemite Valley topographic map at the lower right corner where you see the latitude value for the southern boundary of the map and the longitude value for the eastern border of the map. Each corner has similar latitude and longitude values. The coordinates describe a section of the earth's surface. When the section is rectangular, as most topographic maps are, it is called a quadrangle. The Yosemite Valley map is a special topographic map designed to focus on the valley and is not a quadrangle.

The top of a topographic map is always north. The top and bottom boundaries are lines of latitude; the right and left boundaries are always lines of longitude. The actual latitude or longitude of each boundary are given at the corners. It is sometimes confusing to figure out which is which. An easy way to determine whether a value is latitude or longitude is that latitude readings can never exceed 90° and the longitude values *in the United States* are always greater than 90°.

Using the coordinates at the lower right corner, place a small "X" on the globe in **Figure 1.1** approximating the position of Yosemite National Park. Remember that all latitude coordinates in North America are N and all longitude coordinates are W. This is why coordinates on U.S. topographic maps usually contain numbers only, without the N for north latitude and W for west longitude.

Quadrangle maps are published in several sizes, but two of the most common are 15-minute quadrangle maps and 7.5-minute quadrangle maps; see **Figure 1.1** for relative sizes.

Now take another look at the coordinates of your Yosemite map. What is the *maximum* latitude and longitude range of the Yosemite Valley map?

_____ maximum latitude range

_____ maximum longitude range

Compass Bearings

One frequent task done using a topographic map is to determine the distance and direction of one feature relative to another. For instance, a geologist might want to know how far away a mine is from his current location and in what direction. The direction information is determined by using the map to take a compass bearing.

A bearing is the compass direction of one point relative to another. Bearing is expressed relative to true north or true south. True north and true south are the lines between a point on Earth's surface and Earth's North Pole or South Pole, respectively. They are different than magnetic north and magnetic south, the directions to which a magnetic compass needle points. (The locations of magnetic north and magnetic south are not fixed, but vary from year to year. For instance, magnetic north is currently about 450 miles northwest of Hudson Bay in Northern Canada, and is moving slowly but continuously.)

Bearing is expressed in one of two ways:
- A quadrant bearing expresses direction in degrees east or west of true north or true south.
- An azimuth bearing expresses direction in degrees between 0° and 360° where north is 0 degrees or 360 degrees, east is 90 degrees, south is 180 degrees, and west is 270 degrees.

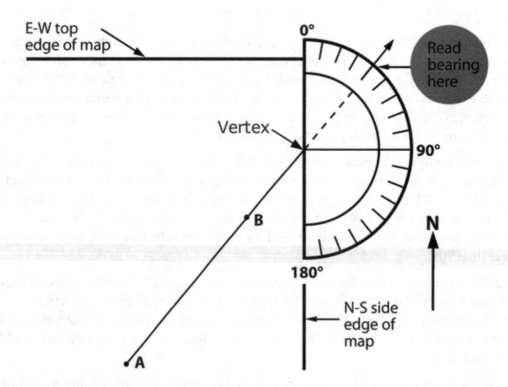

Figure 1.2 Reading Bearings with a Protractor. Illustration by Don Vierstra.

Figure 1.2 illustrates how to determine a quadrant bearing from a starting point, A, on a map to a destination point, B, using a protractor:

(1) Draw a line through the two points so that it extends beyond the map border, either the left (west) edge or the right (east) edge. (See below for instructions on those occasions that the line would not pass through one of these edges)

(2) Orient a protractor so that its 0° and 180° marks are on or parallel to the right edge (east side) or left edge (west side) of the map with the 0° end toward the top of the map (true north).

(3) Position the vertex (center mark along the bottom) of the protractor on the extended A-B line, keeping the protractor in its north-south orientation.

(4) The quadrant bearing is the value on the protractor where the extended A-B line intersects the angle values. In **Figure 1.2**, the quadrant bearing is 40° east of north, or North 40° East, or N40°E.

If you want to take a quadrant bearing including when the line does not extend through the east or west map edge, you can do so by:

(1) Positioning the protractor so that its straight edge is parallel to the east or west edge of the map, and

(2) Positioning the protractor so that its vertex is at the starting point of the line. You can then determine the quadrant reading as described above.

This method can actually be used for to take any quadrant reading. Aligning the protractor with the edge of the map is recommended because it ensures that its north-south orientation is accurate, but the protractor can be used anywhere on the map as long as its straight edge maintains a north-south orientation.

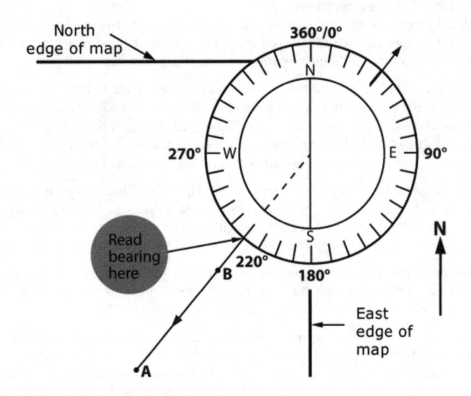

Figure 1.3 Reading Bearings with a Compass. Illustration by Don Vierstra.

Figure 1.3 illustrates how to use a magnetic compass to determine the azimuth bearing of the direction from point B to point A (Note that this is the opposite direction from the quadrant bearing reading in **Figure 1.2**):

(1) As in the above example, draw a line through the two points so that extends beyond the map border, either the left (west) edge or the right (east) edge. Put an arrowhead on the end of the line pointing toward the destination point.

(2) Ignore where the magnetic compass needle is pointing and orient the compass as if it were a 360° protractor; that is, place 0° so it faces the north edge of the map. Then, while maintaining that orientation, position the compass so that the line would go through its center.

(3) The line will intersect with the compass at two points. Choose the reading that corresponds with the direction of the arrow on the line you drew. For instance, if the arrow points toward the southwest (as illustrated in **Figure 1.3**), you would read the azimuth bearing at the southwest intersection of the line with the compass; in this case, 220°.

Note that if you were instead taking a bearing from point A to point B, your arrow would point toward the northeast, and you would read the azimuth bearing at the northeast intersect, which would be 40°.

Public Land Survey System

Latitude and longitude describe the coordinates of a location anywhere on Earth relative to the global baselines of the equator and the prime meridian. In the United States, there exists another grid system for describing a location with the United States.

The U.S. Public Land Survey System (PLSS) is used in most parts of the United States west of the original thirteen states. It describes the location of a parcel of land relative to north-south *principal meridians* and east-west *base lines*. These lines were established by surveyors. Once surveyors established the initial principal meridians and base lines, they surveyed additional east-west parallels and north-south meridians relative to the original two lines.

Figure 1.4 illustrates how the PLSS system works. The strip of land between two parallels is called a township strip and is numbered sequentially away from the base line along with an indicator of whether the township is north or south of the base line; for instance, T1N is Township 1 North, T2S is Township 2 South.

The strip of land between two meridians is called a range strip; like townships, range strips are numbered sequentially away from the principal meridian along with the direction, in this case, east or west. For instance, R1E is Range 1 East, R2W is Range 2 West.

Note that it takes both a township strip and a range strip designation to describe the location of a specific plot of land. The township strip designation describes its location north or south of the base line and the range strip designation describes its location east or west of the principal meridian. Consequently, each plot of land belongs to both a township strip and a range strip.

Each intersection of a township strip with a range strip forms a 6-mile square called a township. Townships are further subdivided as follows:

(1) Each township (6 miles × 6 miles) is subdivided into 36 square-mile squares called sections. Each section contains 640 acres. Sections are numbered from 1 to 36, beginning in the upper right corner, zigzagging back and forth as illustrated in **Figure 1.4**, and ending in the lower right corner of the township square.

(2) Sections are divided into quarters labeled by direction (NE, NW, SE, SW).

(3) Each quarter can be further subdivided into quarters, also with directional labels (NE, NW, SE, SW). Such subdivisions can be done as many times as necessary to describe a location.

Locations are described in terms of this system, *beginning with the smallest subdivision*. For example, point X in **Figure 1.4** is in the southeast one-quarter of the southwest one-quarter of the southeast one-quarter of Section 22 in Township 2 South, Range 3 West, which is written in shorthand as:

$$SE\tfrac{1}{4}, SW\tfrac{1}{4}, SE\tfrac{1}{4}, Sec. 22, T2S, R3W$$

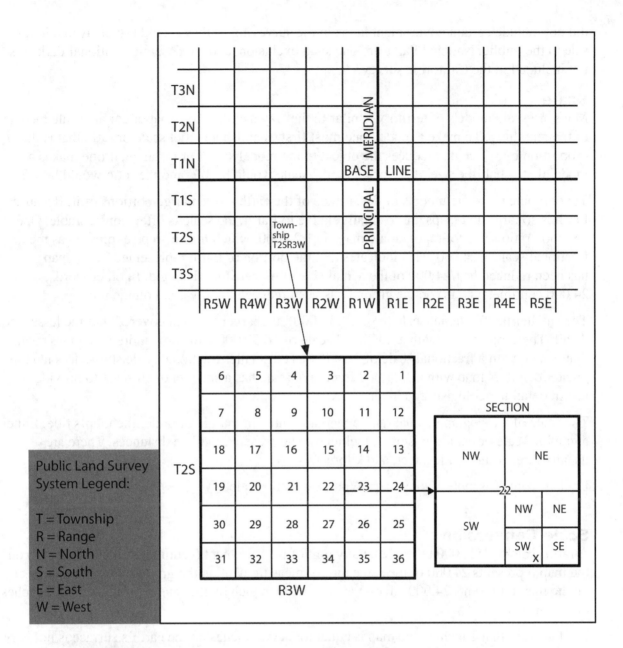

Figure 1.4 The Public Land Survey System. Illustration by Don Vierstra.

Often, the PLSS boundary lines are depicted as a red grid on topographic quadrangle maps. Look at the left side of the Yosemite map. Note the red solid and dashed lines that form the boundaries of squares and partial squares.

> What Township and Range designations describe the area in the southwest (lower left) corner of the Yosemite topographic map?
>
> Township: _____
>
> Range: _____

You may notice that there are no PLSS squares on the north and west side of the map. In this case, the government land was set aside for Yosemite National Park before the California surveying was completed. The Land Ordinance of 1785, which established the survey system, did not include certain government lands in the surveying as they would not be available for sale to the public. National Parks are one such exclusion, and so Yosemite National Park was not included in the California survey.

Scale

Maps are scale models of territory similar to the way a child's toy boat or car are scale models of the real thing. To make a model, one must first establish a model scale or ratio that is the proportion by which one reduces the object to the model size. For instance, if one makes a model that is half the size of the original, the scale would be 1/2, and the ratio would be 1:2.

Topographic maps are models of the surface of the earth—often large portions of it. Because Earth is so large and maps are so small relative to Earth, the scale is often considerable. One common ratio scale for topographic maps is 1:24,000, which may also be expressed as the fractional scale 1/24,000. This indicates that the portion of Earth represented in the map has been reduced to 1/24,000 of the actual size of the territory depicted. In other words, 24,000 centimeters (cm) on the ground has been reduced to only 1 cm on the map.

The smaller the fractional scale (e.g., 1/250,000), the larger the area covered, but the lesser the detail. Therefore, a map with a fractional scale of 1/250,000 will show more area in less detail than a map with a fractional scale of 1/24,000. The smaller fractional scale shows less land in greater detail. A map with a 1/24,000 fractional scale has a small enough scale to provide useful detail to geologists and hikers.

The scale of a topographic quadrangle map is printed in margin or under the map's title. Either a graphic scale or bar scale is often included to help users calculate distances. There are usually three scales: miles, feet, and kilometers.

Look at your Yosemite Valley map. What is its ratio scale? _____

Scale Conversion

A ratio scale of 1:24,000 tells us that any unit of measurement (centimeter, inch, foot, etc.) on the map represents 24,000 of the same unit of measurement on the ground. For example, 1 cm on the map represents 24,000 cm on the ground. One inch on the map represents 24,000 inches on the ground.

But knowing that 1 inch on the map is equal to 24,000 inches on the earth's surface is not very helpful, since no one measures large distances in inches. For the number to be meaningful, it must be converted to a number that is more frequently used to express such distances. This is done by converting the scale from one unit of measurement to another. Since there are 12 inches in a foot, we can divide 24,000 by 12 to calculate that one inch on the map equals 2,000 feet on the ground (24,000 inches × 1 foot/12 inches = 2,000 feet). This type of description, in contrast to a fractional scale, a ratio scale, or graphic scale (like a bar graph), is called a verbal scale.

Scale conversions are important in making use of topographic maps because they provide the reader of the map an idea of the distance between two points on the map in a unit that makes sense. The conversions are done using simple math. Conversion factors appear in Appendix I.

Example 1: One inch on a 1:24,000 scale represents what distance on the ground in miles and kilometers?

Solution:

Using these conversions: 1 foot = 12 inches; 5,280 feet = 1 mile; and 1 mile = 1.609 km,

$$1 \text{ inch on map} = 24,000 \text{ inches on ground} \times \frac{1 \text{ foot}}{12 \text{ inches}} \times \frac{1 \text{ mile}}{5,280 \text{ feet}} = 0.379 \text{ miles}$$

$$0.379 \text{ miles} \times \frac{1.609 \text{ km}}{1 \text{ mile}} = 0.610 \text{ km}$$

Example 2: A mile on the ground would be what distance (in inches) on a 1:24,000 scale topographic map?

Answer:

$$1 \text{ mile on ground} \times \frac{5,280 \text{ feet}}{1 \text{ mile}} \times \frac{12 \text{ inches}}{1 \text{ foot}} \times \frac{1 \text{ inch on map}}{24,000 \text{ feet}} = 2.64 \text{ inches on map}$$

Map Symbols

Topographic maps provide remarkably detailed information about natural and cultural features. The map symbols are located in the USGS Topographic Map Symbols pamphlet in the lab kit, which identifies all the features that may appear on a topographic map. Note that township, range, and section lines are always shown in red. Topographic contour lines are always shown in brown. Rivers, streams, bodies of water, swamps, marshes, and similar features are always shown in shades of blue. And vegetation is always shown in green. **Figure 1.5** illustrates and explains some of the many features found on topographic maps.

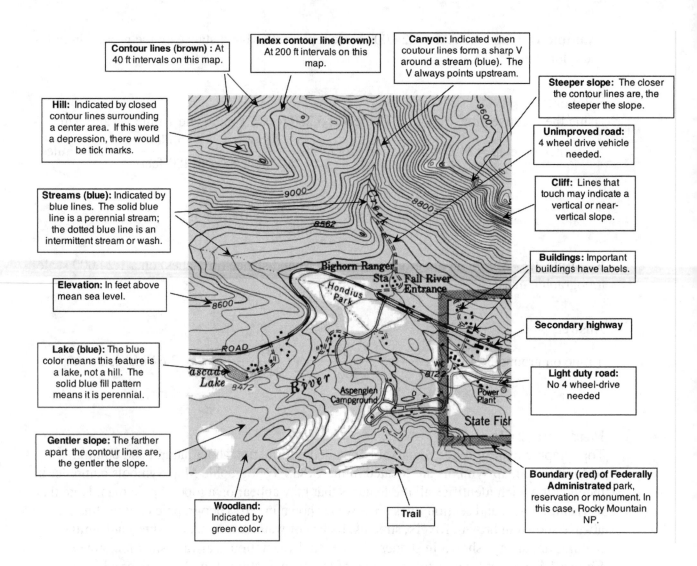

Contour lines (brown) : At 40 ft intervals on this map.

Index contour line (brown): At 200 ft intervals on this map.

Canyon: Indicated when coutour lines form a sharp V around a stream (blue). The V always points upstream.

Steeper slope: The closer the contour lines are, the steeper the slope.

Hill: Indicated by closed contour lines surrounding a center area. If this were a depression, there would be tick marks.

Unimproved road: 4 wheel drive vehicle needed.

Cliff: Lines that touch may indicate a vertical or near-vertical slope.

Streams (blue): Indicated by blue lines. The solid blue line is a perennial stream; the dotted blue line is an intermittent stream or wash.

Buildings: Important buildings have labels.

Elevation: In feet above mean sea level.

Secondary highway

Lake (blue): The blue color means this feature is a lake, not a hill. The solid blue fill pattern means it is perennial.

Light duty road: No 4 wheel-drive needed

Gentler slope: The farther apart the contour lines are, the gentler the slope.

Woodland: Indicated by green color.

Trail

Boundary (red) of Federally Administrated park, reservation or monument. In this case, Rocky Mountain NP.

Figure 1.5 Rocky Mountain National Park. This is a portion of the topographic map of the area near the entrance of Colorado's Rocky Mountain National Park. A color version is available in the online supplement for this lesson. Map: Courtesy of U.S. Geological Survey.

Additional information in the margins of topographic maps includes the revision date, since people are forever changing the man-made features and because the earth's surface changes from events such as earthquakes, landslides, and floods. Some maps feature a purple and red overlay; this is a photo revision produced when aerial photographs are taken to discover new changes in the land surface features. The changes are overprinted on the maps in the standout color of purple and red.

Other information in map margins includes the name of the topographic map, an index map showing the location of the mapped area within a state, the names of adjacent topographic maps, latitude and longitude coordinates of the four corners of the map, and the contour interval. The contour interval is the vertical difference in elevation between the contour lines that show the general shape of the terrain, as described at the beginning of this lesson.

Since the geographic true north (i.e., always at the top of topographic map—north pole) and the magnetic north pole do not coincide, most topographic maps also include in the map margin the magnetic declination, the angle formed between the direction of true geographic north and magnetic north pole. The information is essential to navigation on the ground; a user would need to know magnetic declination to adjust his magnetic compass reading to work with the map. However, the magnetic declination is exact for only the year listed on the topographic map because of the slow migration of the magnetic north pole.

Contour Lines

The most noticeable features of a topographic map are the brown contour lines, also called contours. The brown contour lines illustrate the configuration of the land surface. Every point on a single contour line represents the same elevation. If you could walk along the ground corresponding to a contour line, you would never ascend or descend, you would always be at the same elevation above mean sea level. To aid in counting contour lines on topographic maps, every fifth contour line, called an index contour, is darker brown and is labeled with its elevation above sea level.

The difference in elevation between contour lines is called the contour interval. Another way to think of it is that each contour line represents an increase or decrease in elevation equal to the contour interval. For instance, if the contour interval is 20-feet, each contour line represents either an increase or decrease in elevation of 20 feet.

The contour interval selected by the cartographer (i.e., the map maker) depends on topography. In mountainous terrains, a 100-foot contour interval might be the most suitable because there are so many steep slopes in the mountains. With a 100-foot interval, each contour line represents an increase or decrease of 100 feet in elevation. In the mid-continental area of the United States, a 20-foot contour interval might be the most useful because of the gentle slopes of the hills and valleys in this region. The Mississippi Delta area is so flat that the contour interval is measured in inches. Most USGS topographic maps use the smallest contour interval that is practical because smaller intervals allow for easier readability and provide the detail necessary to identify the geologic features in the area.

Look at your Yosemite Valley map. The contour interval appears under the graphic scale beneath the title.

What is the contour interval on this map? _____

Given this contour interval, what is the change in elevation represented by each contour index line? _____

Also note on your Yosemite map that contour lines tend to be concentric, sharing the same general shape. This is because they illustrate the same topography at different elevations. A radical change in the path of contour lines indicates a radical change in topography. Contours within Yosemite Valley, for instance, follow a different path than the concentric contour lines representing the walls of the valley. In addition, no contour line crosses another. An intersection of contour lines would indicate two different elevations at the same point, something clearly impossible.

Most topographic maps contain benchmarks where elevation and position have been surveyed to the nearest foot. Bench marks are identified on a topographic map by symbols such as "BM" or "X" followed by the elevation. If you were to visit the location represented by a BM symbol, you would find a benchmark, a permanent marker (such as a metal plate embedded in concrete) placed in the field by the USGS or Bureau of Land Management.

Topographic relief is the difference in elevation between two points on a map. Local relief refers to the difference in elevation of adjacent hills and valleys, whereas total relief is the difference in elevation between the highest and lowest points on the map.

Rules for Contour Lines

1. Evenly spaced contour lines depict a uniform slope.

2. Closely spaced contour lines depict a steep slope.

3. Widely spaced contour lines depict a gentle slope.

4. No contour line can cross another contour line.

5. Contour lines may merge to form a single contour line where there is a vertical cliff.

6. Contour lines never split or branch, each contour line is a single, continuous line.

7. Contour lines never end at a point on the map, but may end at the margin of the map.

8. A concentric series of closed contours represents a hill or mountain.

9. A concentric series of closed contours with tick marks (also called hachure marks) represents a closed depression. The tick marks appears on the downhill side.

10. Contour lines form a V pattern when they cross a stream valley or canyon. The apex or point of the V always points upstream (the rule of "Vs").

Online Activities

Per your instructor's directions, go to the online supplement for this lab and complete the activities assigned. Viewing the online videos will help you to complete the quiz.

Lab Exercises

Lab Exercise #1: *Contour Exercises*

Dixie Mine Contour Practice Activity
Prospect Hill Contour Exercise

Dixie Mine Contour Practice Activity

One way to better understand contour lines is to draw them. You will begin this laboratory exercise by learning to draw contour lines on a practice activity. In this exercise, you will draw the contour lines for a set of spot elevations in the Dixie Mine contour activity (**Figure 1.6**), and then check your results against the solution provided at the back of this lab manual and in the online lab supplement.

To begin, watch the video *How to Create Contour Lines for a Topographic Map* in the online supplement, and then follow the steps listed below to draw your own contour map of the Dixie Mine area. Adjustments to your drawing are almost inevitable, so use a pencil. You might also want to photocopy the exercise or download a copy of this exercise from the online supplement for this lab so you can have a fresh copy for each attempt.

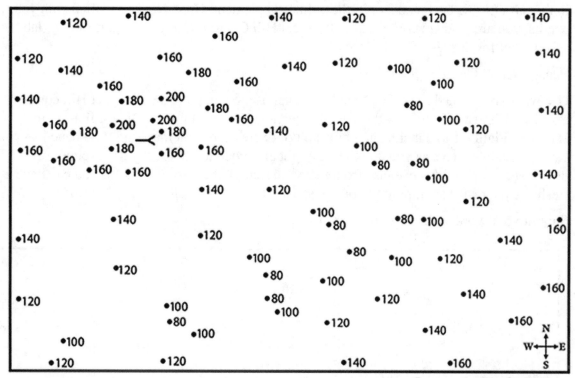

Contour Interval= 20 Feet
—< Tunnel or mine

Figure 1.6 Dixie Mine Contour. Illustration by Don Vierstra.

Instructions and Observations

Step 1: Look at the Dixie Mine contour (**Figure 1.6**) and note areas that are surrounded by lower and upper elevations. Also look for areas where there might be a repeat of a contour line indicating a change in slope direction.

Step 2: Use a contour interval of 20 feet so that each contour line represents a change of 20 feet in elevation above mean sea level. Begin by locating the highest points on the map, and then draw a contour line by connecting the points having that elevation.

Step 3: Find the next highest points on the contour map and connect them, remembering that *contour lines tend to be concentric*. Continue drawing contour lines until you have lines passing through all of the spot elevations. You may need to redraw lines to make them concentric with lines above or below them.

Step 4:   Locate the highest closed concentric contours and write in the label "hill." Locate the lowest closed concentric contours and write in the label "depression." Be sure to add tick marks on the downslope side of the depression contours.

Once you've completed this practice exercise, check your answer with the student answer key in the back of this lab manual or in the online supplement. Once you're satisfied with your results, you are ready to complete the Prospect Hill Contour Exercise as part of your lab assignment for this lesson.

Prospect Hill Contour Exercise

Use what you learned from the Dixie Mine exercise to complete the Prospect Hill contour exercise (**Figure 1.7**). Follow the same instructions you used for the Dixie Mine contour exercise (**Figure 1.6**), labeling any hills and/or depressions you may find. Use pencil so you can make adjustments as you draw. You might also want to photocopy the exercise or download a copy of this exercise from the online supplement for this lab so you can have a fresh copy for each attempt until you are satisfied with the result.

Submit your work as directed by your instructor.

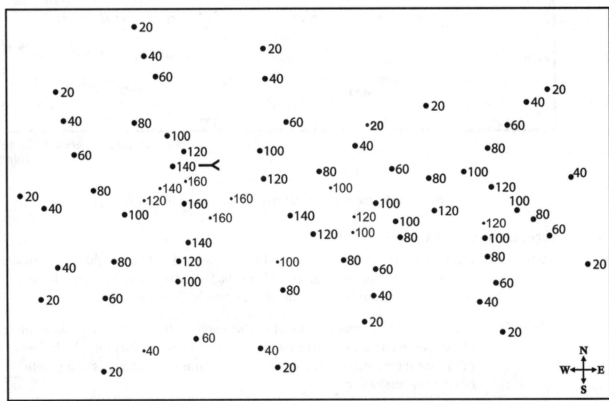

Contour Interval= 20 Feet
≺ Tunnel or mine

Figure 1.7 Prospect Hill Contour. Illustration by Don Vierstra.

Lab Exercise #2: *Topographic Profiles*

Campbell Hill Topographic Profile Practice Activity
Yosemite Valley Topographic Profile Exercise

Topographic maps present an aerial view of the earth's surface, depicting topographic features by means of contour lines and symbols. To view the shape of the earth's surface, however, it is useful to adopt a side-view perspective, as if you were looking at the topography against the horizon. This perspective can be developed from a topographic map by constructing a topographic profile.

As with the contour exercises, you will begin with a practice activity, the Campbell Hill Topographic profile. Use pencil so you can make adjustments as you draw. You might also want to photocopy the exercise or download a copy of this exercise from the online supplement for this lab so you can have a fresh copy for each attempt until you are satisfied with the result.

Campbell Hill Topographic Profile Practice Activity

In this exercise, you will learn how to produce a side view or cross-section of the land surface along line A-A[1] in the contour map in **Figure 1.8a**. Begin by watching the video, *Creating a Topographic Profile of a Land Surface Area*, in the online supplement and then follow the steps listed for this Lab Exercise.

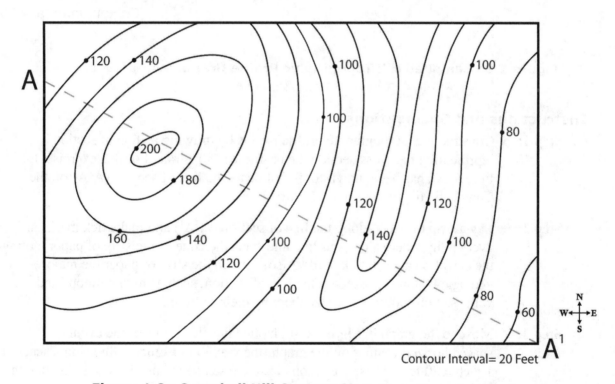

Figure 1.8a Campbell Hill Contour Map. Illustration by Don Vierstra.

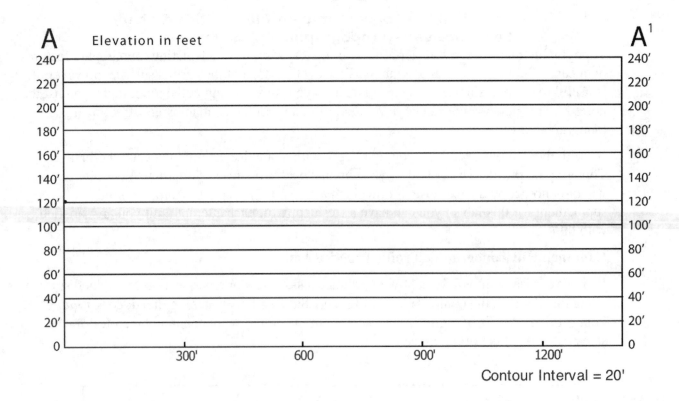

Figure 1.8b Campbell Hill Topographic Profile Graph. Illustration by Don Vierstra.

Instructions and Observations

Step 1: Constructing a topographic profile begins by drawing a "line of section" representing the cross-section the profile will illustrate. For this exercise, the line, A-A[1], has been drawn for you. Locate the line of section (A-A[1]) on the Campbell Hill contour map (**Figure 1.8a**).

Step 2: Lay a strip of paper along the line of section (A-A[1]) and make tick marks at every point where a contour line intersects the edge of the strip of paper, noting the elevation at each tick mark. Also, note on the strip of paper the major geologic features crossed by the line of section, such as mountaintops and valleys. You should recognize them by their contours.

Step 3: Move to the graph in **Figure 1.8b**. Notice the elevations of the evenly spaced horizontal lines on the graph (the *y*-axis) in **Figure 1.8b** using a scale of 1 inch = 80 feet. These elevations were chosen so the profile would include the highest and lowest points along the line of section. Remember also that each horizontal line on the graph represents a constant elevation corresponding to a contour line on the map.

Step 4: Lay the strip of paper along the bottom of the graph. Above each tick mark on the paper strip, place a dot on the graph at the proper elevation.

Step 5: Complete the topographic profile by drawing a smooth line connecting all of the points on the graph.

Step 6: Label the major topographical features (e.g., hill, valley, slope) on the profile.

Note that the topographic profile is an approximation. What is not known from this contour map is the actual elevation of Campbell Hill. We only know that the highest contour line encircling the peak is 200 feet. Likewise, we don't know the elevation of the slope in the lower right hand corner of the map, but we can set a lower limit for this elevation based the contour lines we can see.

What is the lowest elevation to the nearest foot that is possible on this map? _____

Also note that this topographic profile is not exactly like the profile you would see if you saw the same landscape against the horizon. It is vertically exaggerated so that the slopes, valleys, and hills would be more obvious. In other words, In other words, the slopes look steeper than they are in actuality. Note that the horizontal scale of the profile is 1 inch = 300 feet, whereas the vertical scale is 1 inch = 100 feet. The vertical exaggeration is calculated by dividing the horizontal scale by the vertical scale.

Given this information, what is the vertical exaggeration of this topographic profile? _____

Once you're completed this practice exercise and answered the above questions, check your answer against the student answer key in the back of this lab manual or by downloading the solution from the online lab supplement. Once you're satisfied with your results, you are ready to complete the Yosemite Valley Topographic Profile as part of your lab assignment for this lesson.

Yosemite Valley Topographic Profile Exercise

Use what you learned from the Campbell Hill practice activity to complete the Yosemite Valley Topographic Profile (**Figure 1.9a**) using the contour map below. Note that this contour map uses a 500-foot contour interval rather than the 40-foot interval on the map provided with your lap kit. Follow the same instructions as for the Campbell Hill Topographic Profile. Find and label the same peaks and creek seen on the contour map (**Figure 1.9a**).

Use pencil so you can make adjustments as you draw. You might also want to photocopy the exercise or download a copy of this exercise from the online supplement for this lab so you can have a fresh copy for each attempt until you are satisfied with the result.

Submit your work as directed by your instructor.

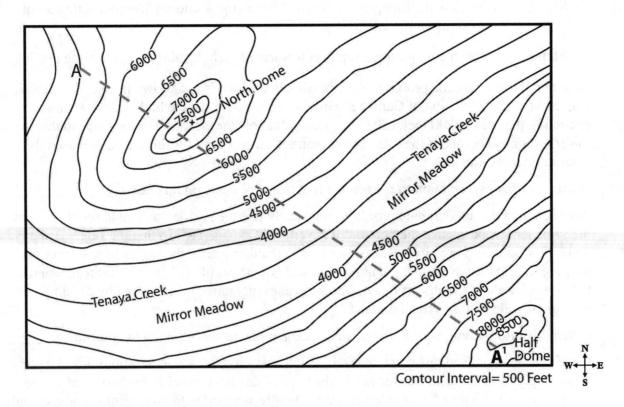

Figure 1.9a Yosemite Valley Contour Map. Illustration by Don Vierstra.

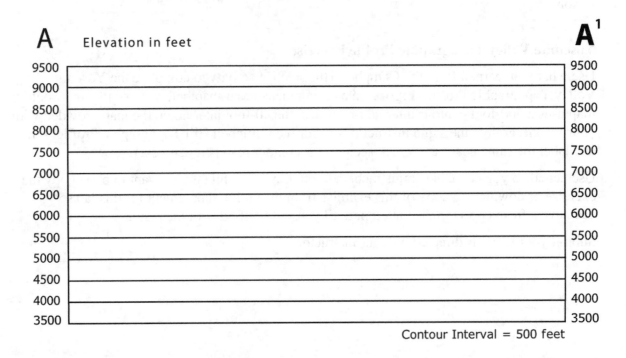

Figure 1.9b Yosemite Valley Topographic Profile Graph. Illustration by Don Vierstra.

Quiz

Multiple Choice

Questions 1 and 2 are based on **Lab Exercise #1:** *Contour Exercises.*

1. According to the Prospect Hill contour map in **Figure 1.7**, the land surface could be
   a. a hill.
   b. two valleys with a hill in the middle.
   c. two hills with a lower area between them.
   d. two closed depressions with a flat-topped plateau in the middle.

2. In **Figure 1.7**, what is the elevation of the mine (locate the mining symbol) on Prospect Hill?
   a. 100 feet
   b. 120 feet
   c. 140 feet
   d. 150 feet

Question 3 is based on **Lab Exercise #2:** *Topographic Profiles.*

3. What answer below best represents the profile in **Figure 1.9a** between point A and point A^1?

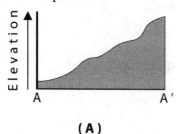

(A)

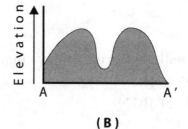

(B)

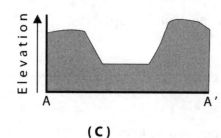

(C)

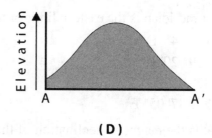

(D)

For questions 4 and 5, you will want to refer to information found in the Scale Conversion section of this lesson. Round your answers to the nearest hundredth decimal place.

4. On a 1:125,000-scale map, one inch represents what distance on the ground?
   a. 1,250 feet
   b. 1,970 km
   c. 1.97 miles
   d. 1.25 miles

5. A mile on the ground would be what distance (in inches) on a 1:50,000 scale topographic map?

    a. 0.79 inches

    b. 1.26 inches

    c. 1.05 inches

    d. 5,000 inches

Questions 6 through 28 are based on the topographic map of Yosemite Valley, Yosemite National Park, California, and the USGS Topographic Map Symbols pamphlet (both found in the lab kit).

6. What is the general location of this map area in the state of California?

    a. east central California

    b. north west California

    c. north east California

    d. south central California

7. Using the USGS Topographic Map Symbols pamphlet in the laboratory kit, what type of lake is Starr King Lake on the USGS Yosemite Valley topographic map?

    a. annual lake

    b. perennial lake

    c. intermittent lake

    d. dry lake

8. What parallel marks the northern limit of the topographic quadrangle map?

    a. 119° 45' N

    b. 119° 29' 10" N

    c. 37° 42' N

    d. 37° 47' 05" N

9. What meridian is the eastern limit of the topographic quadrangle map?

    a. 119° 45' E

    b 119° 29' 10" W

    c. 37° 42' W

    e. 37° 47' 05" E

10. What is the magnetic declination of the topographic quadrangle map?

    a. 10.5° east

    b. 15.5° west

    c. 17.5° east

    d. 20.5° west

11. In what year was the magnetic declination measured?

    a. 1927

    b. 1955

    c. 1958

    d. 1970

12. This topographic quadrangle map was compiled from aerial photographs taken in
    a. 1927.
    b. 1955.
    c. 1958.
    d. 1970.

13. What is the fractional scale of the topographic map?
    a. 1/12,000
    b. 1/24,000
    c. 1/62,500
    d. 1/250,000

14. How can the ratio scale of this map be expressed as a verbal scale?
    a. 1 inch = 1,000 feet
    b. 1 inch = 2,000 feet
    c. 1 inch = 1 mile
    d. 1 inch = 4 miles

15. Using the bar scale provided in the map margin, measure the length of the southern boundary of the topographic map from El Portal to Merced Peak. The distance in miles is about
    a. 5 miles.
    b. 9 miles.
    c. 14.5 miles.
    d. 18.5 miles.

16. Why are there so many curves and switchbacks in the Glacier Point Road?
    a. The slope is relatively flat and the curves give people more time to look at the scenery.
    b. The slope is steep and the curves force traffic to move at a slower pace.
    c. The slope is steep and the curves make the road safer and easier to drive.
    d. The curves keep vehicles from the edges of the cliffs.

17. What direction does the Merced River flow through the valley floor from Sentinel bridge to El Capitan Bridge?
    a. southwest
    b. northeast
    c. southeast
    d. northwest

18. Starr Lake drains in which direction?
    a. northwest
    b. northeast
    c. southwest
    d. southeast

19. Which of the following areas of Yosemite Valley appears to be the steepest?
    a. Inspiration Point
    b. Rockslides
    c. Ribbon Meadow
    d. Middle Brother

20. Identify the section, township, and range for Elephant Rock.
    a. Sec. 36, T.2S, R20E
    b. Sec. 35, T.2S, R21E
    c. Sec. 2, T.3S, R20E
    d. Sec. 2, T.3S, R21E

21. What is the contour interval of the USGS topographic map of Yosemite Valley in your lab kit?
    a. 20 feet
    b. 40 feet
    c. 100 feet
    d. 200 feet

22. If you were a casual recreational hiker with young children, which of the following trails would be least strenuous for a leisurely family outing?
    a. Merced Lake Trail
    b. Mirror Lake Loop Trail.
    c. Tenaya Lake and Tuolumne Meadows Trail
    d. El Capitan Trail

23. What mountain, peak or dome has the highest elevation point on the topographic map?
    a. Half Dome
    b. Clouds Rest
    c. Basket Dome
    d. Sentinel Dome

24. What is the elevation of the lowest point benchmark "BM" on the topographic quadrangle?
    a. 3,182 feet above sea level
    b. 3,227 feet above sea level
    c. 3,419 feet above sea level
    d. 3,438 feet above sea level

25. What is the total relief within the topographic quadrangle map?
    a. 5,615 feet
    b. 6,018 feet
    c. 6,507 feet
    d. 6,699 feet

26. The approximate shortest road miles distance from Yosemite Lodge to Mirror Lake is _____ miles. (Use your ruler or protractor to measure and calculate the approximate distance in miles. Remember 1 inch equals 2,000 feet.)

   a. 1 mile

   b. 3 miles

   c. 5 miles

   d. 6 miles

27. How long is the Wawona Tunnel just north of Inspiration Point on Wawona Road?

   a. 3850 feet

   b. 4250 feet

   c. 4600 feet

   d. 4800 feet

28. What is the quadrant bearing from the first "L" in Lost Lake to the "Q" in Quarter Domes near the east end of Yosemite Valley topographic map?

   a. 35°

   b. N35°E

   c. 45°

   d. S45°E

Questions 29 and 30 refer to "The Geological Story of Yosemite Valley," which can be found on the back of the USGS Yosemite Valley topographic map.

29. What type of rock did the glaciers and Merced River excavate to form Yosemite Valley?

   a. granite

   b. sandstone

   c. shale

   d. quartzite

30. What is the shape of the Yosemite Valley, which is typical of glaciated valleys?

   a. V shape

   b. U shape

   c. W shape

   d. S shape

Short Answer

1. Why are the N, S, E, and W designations often omitted from coordinates on U.S. topographic maps?

2. What is the difference between a quadrant compass bearing and an azimuth compass bearing?

3. How small can the divisions of a PLSS section be?

4. Why can't contour lines ever cross?

5. Why don't all topographic maps have the same contour interval?

6. What do the colors blue and green indicate on a topographic map?

7. Why would a topographic map be updated?

Student Answer Key

Pages 4–5:

Practice Activity on Topographic Maps

What is the latitude and longitude value of Point A?

Latitude: 20° N
Longitude: 120° W

What about Point B?

Latitude: 40° S
Longitude: 20° W

Using the coordinates at the lower right corner, place a small "X" on the globe approximating the position of Yosemite National Park on Figure 1.1.

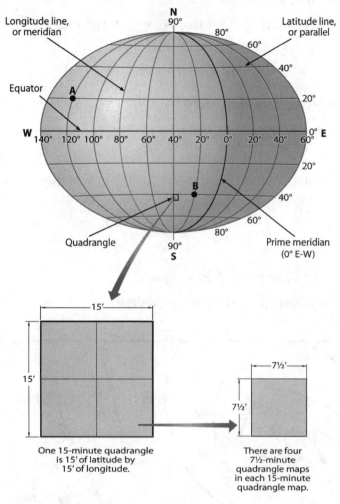

What is the maximum latitude and longitude range of the Yosemite Valley map?

5'5" maximum latitude range (the difference between the top and bottom coordinates on the east side of the map)

15'50" maximum longitude range (the difference in coordinates between the east and west sides of the map)

Page 9:

Public Land Survey Practice Activity

What Township and Range designations describe the area in the southwest (lower left) corner of the Yosemite topographic map?

Township: T. 3. S.
Range: R. 20. E

Page 10:

Scale Practice Activity

Look at your Yosemite Valley map. What is its ratio scale?

1:24,000

Page 13:

Contour Lines Practice Activity

What is the contour interval on this map?

40 feet

Given this contour interval, what is the change in elevation represented by each contour index line?

200 feet

Page 15:

Lab Exercise #1: *Contour Exercises*

Dixie Mine Contour Practice Activity

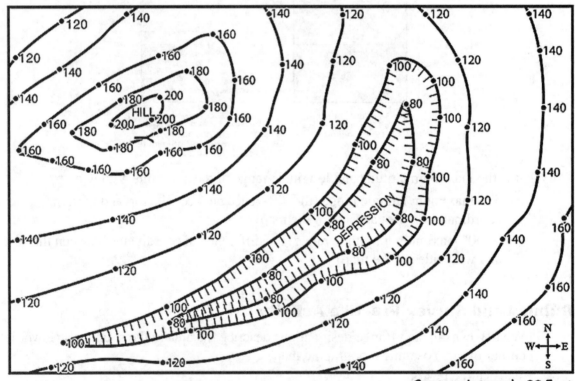

Contour Interval= 20 Feet
⌐< Tunnel or mine

Lab Exercise #2: *Topographic Profiles*
Campbell Hill Topographical Profile Practice Activity

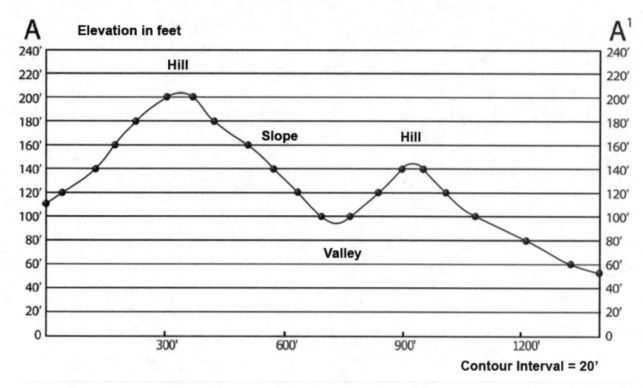

What is the lowest elevation that is possible on this map?
    41 feet

What is the vertical exaggeration of this topographic profile?
    3X

PLATE TECTONICS

Lesson 2

AT A GLANCE

Purpose

Learning Objectives

Materials Needed

Overview

History of Plate Tectonics
Seafloor Exploration
Magnetism and Paleomagnetism
The Theory of Seafloor Spreading
Earth's Interior
Principles of Plate Tectonics
Plate Boundaries

Lab Exercise

Lab Exercise: *Plate Boundaries*

Online Activities

Quiz

Multiple Choice
Short Answer

Purpose

This laboratory lesson will provide you with a foundation and understanding of plate tectonics.

Learning Objectives

After completing this laboratory lesson, you will be able to:

- Explain the hypothesis of continental drift.
- Explain how Earth's magnetism, paleomagnetism, and magnetic reversals aided development of the theory of plate tectonics.
- Describe the layers composing Earth's interior.
- Describe how tectonic plates move and how they interact at the three basic types of plate boundaries.
- Explain how the formation of volcanic islands at hot spots provides evidence in support of plate tectonic theory.

Materials Needed

- ❏ Pencil
- ❏ Eraser

Overview

For centuries, scientists and laypeople alike wondered where mountains came from, why there were continents and oceans, why earthquakes shook the ground, and why volcanoes erupted. Many ideas were offered, some of them supernatural and some more scientific. But the idea that finally explained all of the above was plate tectonics theory.

Plate tectonics theory is a relatively recent development that revolutionized the science of geology. As discussed below, the theory holds that Earth's rigid outer layer, called the lithosphere, is broken into large moving sections or plates. Many of the features found at plate edges, such as folded mountain ranges, volcanoes, trenches, rift valleys, and island arcs have been created by plate tectonic activity.

History of Plate Tectonics

Ever since cartographers produced maps with a sufficient degree of accuracy to portray the shapes of the continents realistically, individuals have noticed the matching coastlines on opposite sides of the Atlantic Ocean. As early as 1620, the English philosopher, statesman, and scientist Francis Bacon commented on the similarity of the Atlantic coastlines of South America and Africa. The similarity had caused him and others to wonder whether the two continents had once been joined like two pieces of a giant jigsaw puzzle or if the apparent match was just a coincidence.

Most scientists at the time were convinced that it was indeed just a coincidence, but one prolific geologist, Eduard Suess, made a different suggestion. In the early 1800s, Suess noted not only the fit between the continents but also the discovery of identical *Glossopteris* seed fern fossils in South America, Africa, and Australia. He wondered how the same plants had ended up on different continents. The only explanation that made sense to him was that all the southern continents, including Antarctica, had once been joined into one larger continent that he called Gondwana or Gondwanaland (**Figure 2.1**). Suess' ideas were not at all taken seriously and were discarded. However, in the early 1900s, several geologists took a second look, the most noteworthy of whom was the German geophysicist and meteorologist Alfred Wegener.

Wegener went further than Suess, both in his hypothesis and in the evidence he gathered to develop it. He noted other identical sets of fossils that appeared on different continents separated by oceans (**Figure 2.1**), and wondered by what means these animals might have traveled overseas. He observed that striations, long scratches in rock left behind by glaciers, appeared in areas that are now tropical.

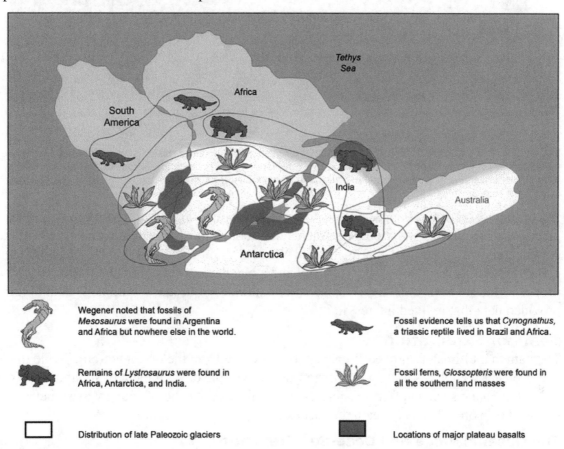

Figure 2.1 Gondwana. Various kinds of data exist to support the idea that the southern continents were once joined into a supercontinent some 200 million years ago. From *Planet Earth* by John J. Renton. Copyright © 2002 by John J. Renton. Reprinted by permission of Kendall Hunt Publishing Company.

Wegener proposed not only that South America and Africa had once been joined but also that the northern continents had been joined together to form a larger continent, Laurasia. In fact, his evidence seemed to indicate that *all* of the continents had once been connected together into one supercontinent he called Pangaea (**Figure 2.2**).

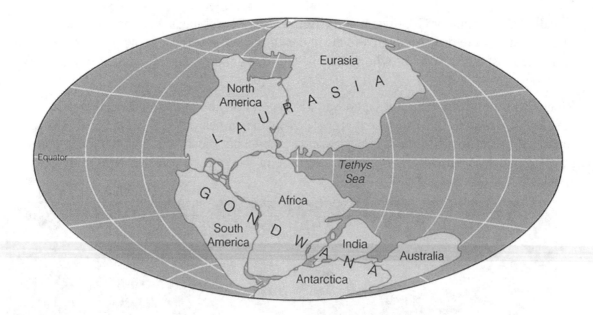

Figure 2.2 Supercontinent. Alfred Wegener proposed that all of the present continents were once joined into a supercontinent he called Pangaea. From *Planet Earth* by John J. Renton. Copyright © 2002 by John J. Renton. Reprinted by permission of Kendall Hunt Publishing Company.

According to Wegener, Pangaea began to break up about 200 million years ago, with the pieces drifting apart as individual continents into today's positions, a movement referred to as continental drift. The major shortcoming of Wegener's proposal was that he could not identify a scientifically defensible source of energy and a mechanism for continental drift. His proposal was set aside and largely ignored for a number of decades, until new technologies provided data that revived interest in it.

Seafloor Exploration

The data that ultimately proved Wegener correct came from the ocean bottom. While there had been early attempts to map the ocean floor, it was the development of sonar after World War I that made such an effort feasible. Sonar analyzes the echo of sound waves that are bounced off solid objects beneath the surface, such as the ocean floor.

The Oceanic Ridge and Deep-Sea Trenches

The first major feature to be discovered on the ocean floor was the oceanic ridge, an undersea mountain range extending 65,000 km (40,000 miles) through every ocean on Earth. The oceanic ridge is such a dominant feature that if one were approaching Earth from space and the ocean basins were empty, the oceanic ridge would be the first surface feature to be identified, even before major mountain ranges such as the Rockies, the Alps, and the Himalayan Ranges.

This extensive feature turned out to be more than a mountain range, however. Further investigation revealed that at the center of the oceanic ridge is a rift, a break in the ocean floor, through which magma, or molten rock, is continually rising and solidifying into basalt, the rock that composes oceanic crust, or the rocks that compose the top layer of the sea bottom.

Another major feature discovered on ocean bottoms was a number of long, narrow, deep trenches—referred to as deep-sea trenches—that paralleled some continental coasts and island chains. The trenches are the deepest spots in the ocean. The deepest trench is the Mariana

Trench that extends for 2550 km (1560 miles) east of the Philippine Islands, China, and Japan; the trench is more than seven miles deep.

The oceanic ridge and deep-sea trenches were major discoveries, the importance of which would be realized only after additional new technology analyzed the magnetism of seafloor rocks.

Magnetism and Paleomagnetism

Earth is a planet with a magnetic field having both north and south poles. As discussed in Lesson 1, magnetic north is not in the same location as true north. Earth's magnetic field extends from its center with an axis inclined at approximately 11.3 degrees to Earth's rotational axis (**Figure 2.3**). The north and south magnetic poles are not static but wander about 15 km (9 miles) per year. For instance, in 1838, the north magnetic pole was located on the west coast of the Boothia Peninsula in the Arctic (95° West longitude. and 70° North latitude); it is now about 450 miles northwest of Hudson Bay in Northern Canada.

Figure 2.3 Earth's Magnetic Field. Earth's magnetic field is similar to having a slightly tilted bar magnet in its interior. A "wobble" in the tilt causes the locations of north and south magnetic poles to vary slightly from year to year. Illustration by Don Vierstra.

Not only does the orientation of Earth's magnetic field change, but those changes are constantly being recorded within newly formed basalt that forms at the rifts of the oceanic ridges. Basalt contains large amounts of iron in the form of the mineral magnetite, named for its magnetic properties. As basaltic lava (as magma is called once it reaches the surface) cools, the new magnetite grains form tiny permanent magnets that have the same orientation as Earth's magnetic field, capturing that orientation at the moment of their formation. This is called magnetic alignment.

Thus, the orientation of Earth's magnetic field is frozen in rocks containing such magnetic minerals. The basalt of the ocean bottom has large amounts of iron, but there are also some rocks in the continental lithosphere that contain magnetic minerals. A magnetometer is an instrument that can detect the direction of that magnetic alignment. This new technology allows geologists to determine the orientation of the Earth's magnetic field at the time the rock formed, a capability that has launched the field of paleomagnetism, the study of the changes that have occurred in Earth's magnetic field. By determining the orientation of the magnetic field within lava flows of different ages, the history of Earth's magnetic field over the period of time represented by the lava flows can be determined.

Magnetic Reversals

Earth's magnetic history includes not only the movement of magnetic north and south, but times during which Earth's magnetic field appears to have undergone reversals—that is, the north and south magnetic poles switched locations. Which pole is north and which pole is south is called magnetic polarity. Today, the north magnetic pole is in the Arctic region, while the south magnetic pole is in the Antarctic region. This is called normal polarity. Reverse polarity occurs when poles switch places so that a compass points north to the Antarctic and south to the Arctic.

The discovery, that Earth has magnetic reversals was made by a Japanese scientist, Motonori Matuyama, in 1920. Since Matuyama's original discovery, scientists have learned that magnetic reversals have occurred at an average rate of 4 to 5 reversals per million years over the past 10 million years, with the last reversal occurring about 780,000 years ago. With all that we know about Earth magnetism and magnetic reversals, there is still no agreement as to what causes reversals to take place.

Every time a magnetic reversal occurred, the basalt that formed after the reversal recorded the new polarity with the magnetite crystals pointing to the new North Pole.

The Theory of Seafloor Spreading

The collection of paleomagnetic data from ocean surveys increased after World War II. Greatly puzzling to geologists was the presence of magnetic striping—bands of rocks with opposite magnetic polarity—that appeared on each side of the oceanic rift zones (**Figure 2.4**).

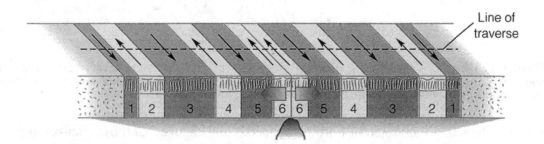

Figure 2.4 Magnetic Striping. Paleomagnetic data gathered during a traverse, or crossing, of an oceanic ridge revealed magnetic striping of the seafloor. From *Planet Earth* by John J. Renton. Copyright © 2002 by John J. Renton. Reprinted by permission of Kendall Hunt Publishing Company.

The explanation lay in a proposal made by a Princeton University geologist, Harry Hess. Hess had proposed that the rocks of the ocean floor were constantly spreading away from the oceanic ridge. Hess asserted that outflow of lava at the rift zones continually forms new oceanic crust. As magma wells up into the rift zone it creates pressure that moves the crust away from the rift in two opposing directions. The addition of the new crust to existing crust actually widens the seafloor, a concept called seafloor spreading (**Figure 2.5**).

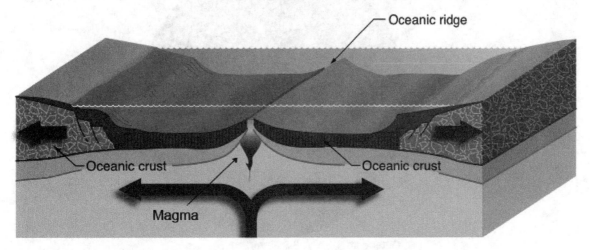

Figure 2.5 Seafloor Spreading. Magma makes its way to the surface where new oceanic crust is being formed. In doing so, newly formed rock is moving the oceanic crust in opposite directions from the mid-oceanic ridge. Illustration © Kendall Hunt Publishing Company.

As newly erupted magma solidified into new crust, it recorded Earth's magnetic field, and then moved away from the rift as more rock formed at the rift. When a magnetic reversal occurred, the basalts that formed after the reversal recorded the altered magnetic field, with the iron crystals in that rock pointing to the new North Pole. The longer the time between magnetic reversals, the wider the magnetic stripe that was created. Hess conjectured that if seafloor spreading were a fact, there should be two such patterns of magnetic striping, one on each side of the rift, that are mirror images of each other.

When paleomagnetic data confirmed that there are, indeed, mirror images of magnetic striping along the oceanic rifts zones, it turned the geological world upside down. Seafloor spreading was validated as a possible mechanism for Alfred Wegener's continental drift hypothesis. The explanation became plate tectonics theory.

Earth's Interior

To understand plate tectonics theory, it's necessary to take a look at Earth's interior. While no one has traveled more than a few miles inside Earth, scientists have found ways to "see" inside our planet through their analysis of seismic waves, the energy released by earthquakes transmitted to far away locations through rock movement. They have discovered that Earth is composed of concentric layers of rock, each layer hotter and denser than the one above it (see **Figure 2.6**).

- The crust is Earth's thin, outermost layer.
- The mantle, Earth's thickest layer, lies below the crust. Mantle rock is hot enough to behave more like Silly Putty® than a solid.
- Below the mantle, about 1800 miles below the surface, is the outer core, which is so hot that the rock is liquid.
- At the very center of the earth is the inner core, a region where pressures are so intense that the iron and nickel composing it are solid, even though the heat is well above the melting point for these metals, some 6700°F.

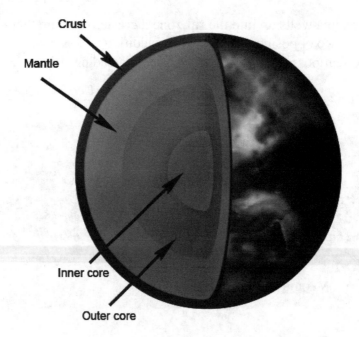

Crust

Mantle

Inner core

Outer core

Figure 2.6 Earth's Interior. Earth's interior is composed of a brittle crust, solid mantle, liquid outer core, and solid inner core. Illustration: Shutterstock 3618258, credit Lorelyn Medina.

These layers are based on the composition of the rocks composing them. The lighter, less dense rocks are in the crust, whereas the heaviest, most dense rocks are in the inner core.

Geologists also categorize Earth's layers based on the properties of the rock. Temperatures within Earth rise with depth, something scientists call the geothermal gradient. Rocks in Earth's crust and the upper part of the mantle are relatively cool and brittle. This cool, brittle layer is called the lithosphere (**Figure 2.7**).

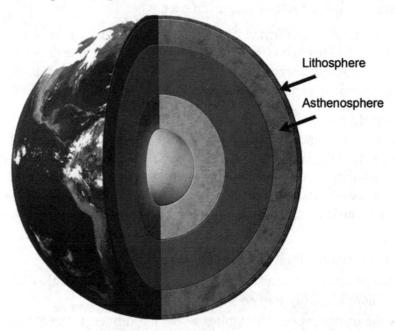

Lithosphere

Asthenosphere

Figure 2.7 Lithosphere and Asthenosphere. The properties of rock within Earth vary by layer. The crust and top part of the mantle together form the lithosphere; it is less dense and brittle. The mantle beneath the lithosphere, called the asthenosphere, is denser than the lithosphere and is plastic rather than brittle. The difference in density causes the brittle lithosphere to float on the plastic asthenosphere. Illustration: Shutterstock 11968393, credit Andrea Danti.

As temperatures and pressures rise with depth, according to the geothermal gradient, the hot rocks begin to behave more like a plastic. They become malleable and will even flow. A good example of this behavior is Silly Putty® or warm taffy. The layer of rocks bearing these properties is called the asthenosphere.

The distinction between the lithosphere and asthenosphere is key to understanding plate tectonics. The lithosphere is generally less dense than the asthenosphere; that is, its rocks either contain lighter atoms or are less closely packed together. We generally think of density in terms of weight: less dense materials weigh less than more dense materials. Density is measured in terms of water, which weighs 1.0 gram per cubic centimeter (g/cm^3).

Less dense materials float on a denser fluid; for instance, an ice cube floats in water because it is less dense than the water. In this case, the difference in density exerts an upward force known as buoyancy. The relative weight of substances, based on their density, determines how deeply they will sink. An ice cube has a density of 0.917 g/cm^3 so 91.7 percent of it sinks below the water's surface. Pinewood has density of 0.530 g/cm^3 so only 53 percent of it sinks into the water. The point of balance between the downward pull of gravity and the upward force of buoyancy is called gravitational balance.

Likewise, the lithosphere and asthenosphere also exist in gravitational balance, with the lithosphere floating on the plastic asthenosphere (**Figure 2.8**).

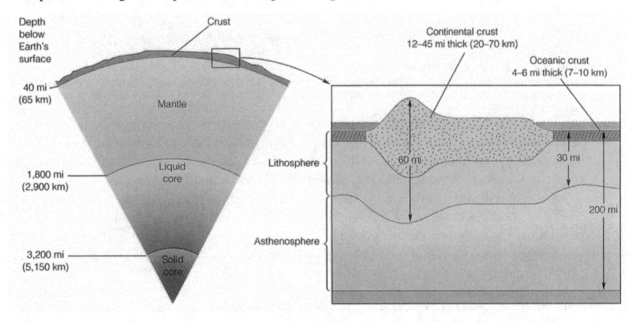

Figure 2.8 Earth's Interior. The three major subdivisions of Earth are the core, mantle, and crust. The crust and the outer brittle portion of the mantle are combined in the lithosphere, which in turn overlies a plastic portion of the mantle called the asthenosphere. From *Planet Earth* by John J. Renton. Copyright © 2002 by John J. Renton. Reprinted by permission of Kendall Hunt Publishing Company.

Principles of Plate Tectonics

Refining plate tectonics theory owes much to the development of another new technology, tomography, which is capable of producing images of what lies beneath Earth's surface. Tomography revealed that the lithosphere is not one solid mass; it is broken up into rigid plates, each of which floats on the asthenosphere. The speed of plate movement varies, but is roughly the same speed at which fingernails grow—about 1 to 5 cm per year.

There are twelve major plates, plus a number of minor plates, each of which is composed of oceanic and continental lithosphere moving over the asthenosphere and following its own path. Thus, continents do not move through the oceans; it is the lithospheric plates that move, carrying the continents with them. **Figure 2.9** shows the major plates and the direction of their movement.

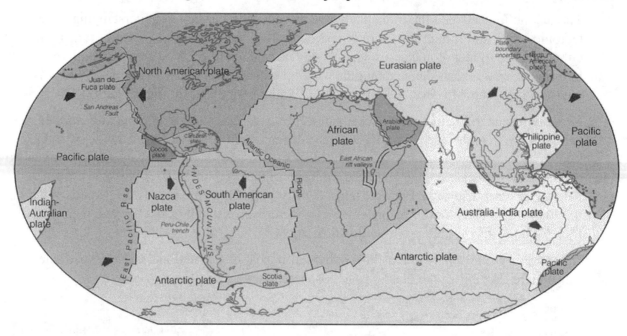

Figure 2.9 Plate Boundaries. The lithosphere is broken into about a dozen large plates and many smaller plates. The number of plates change with time as large plates break apart and smaller plates collide and weld together. Sawtooth lines indicate a convergent boundary, the solid lines indicate divergent boundaries, and transform boundaries are shown with thinner solid lines. From *Planet Earth* by John J. Renton. Copyright © 2002 by John J. Renton. Reprinted by permission of Kendall Hunt Publishing Company.

But what makes the plates move? One proposed mechanism is a process called convection. Convection is a circular movement that occurs when a substance is heated and subsequently cooled. A warmer substance has less density. So if it is fluid, like air or water—or, in this case, magma—it will rise away from the source of the heat. Once away from the heat, the substance will cool, become denser, and sink. As it moves, other fluid moves in to take its place and undergoes the same heating and cooling. The process of rising and sinking causes a circular current called a convection cell. The lava that extrudes at oceanic ridges rises to the surface as upwelling magma at the rising part of a convection cell (**Figure 2.10**).

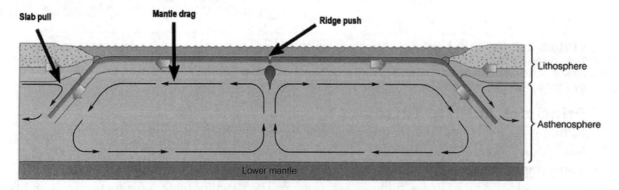

Figure 2.10 Mantle Convection Cells. Heat from Earth's inner core and geothermal gradient heats mantle rock to form convection currents that may be one of the forces behind tectonic plate movement. Illustration by John J. Renton.

Lesson 2/Plate Tectonics

The heat necessary to create convection within Earth's mantle has two sources:

(1) Earth's liquid outer core

(2) The geothermal gradient (the deeper into the Earth one goes, the hotter the temperature)

Convection cells are thought to be located within the plastic asthenosphere, where some geologists think they may supply the energy and one mechanism to move tectonic plates—mantle drag. With mantle drag, the friction of moving asthenosphere at the point of contact with the lithosphere would drag the lithospheric plate along.

Other proposed mechanisms are ridge push and slab pull. Ridge push is a twofold force that occurs when new oceanic crust is formed at the rift zone. First, the creation of new seafloor would push the older seafloor away from the rift. Second, the new seafloor rides high on oceanic ridge where gravity would cause the rock to slide downhill, pushing the older oceanic crust ahead of it. Slab pull also depends on gravity; in this case, the sinking lithospheric plate at a subduction zone would pull the younger plate behind it.

Mantle drag, ridge push, and slab pull are possible explanations for the movement of tectonic plates, each of which moves along its own path. The dividing line between two plates is called a plate boundary.

Plate Boundaries

There are three types of plate boundaries (see **Figure 2.9**): (1) divergent boundaries, where two plates are moving apart from each other, (2) convergent boundaries, where the plates are moving toward each other, and (3) transform boundaries, where the plates are sliding past each other. Each type of boundary has distinctive features associated with it.

Divergent Plate Boundaries

Divergent plate boundaries are those where two adjacent plates are moving apart. The rift zone from which lava extrudes to form new oceanic crust is an example of a divergent plate boundary. If the area above the rising portion of a convection cell is a plate boundary, the two plates move apart. This is the case at the oceanic ridges; the moving plates have the effect of widening the ocean basin, creating an opening ocean.

If the area above the rising portion of a convection cell is the interior of a tectonic plate, the convection cell causes stretching forces in the lithosphere. Over time, the stretching forces may cause the lithosphere to rift, or break, slowly coming apart. The break is called a rift valley, the largest one of which is in East Africa. As the rift valley grows, it may eventually reach the ocean and the ocean will extend into the valley. This is called a linear ocean; the Red Sea is the best example of a linear ocean on today's Earth (**Figure 2.11**). Eventually, the rifting will become complete, the plates separate, and the area becomes a divergent plate boundary. This is how Pangaea broke up.

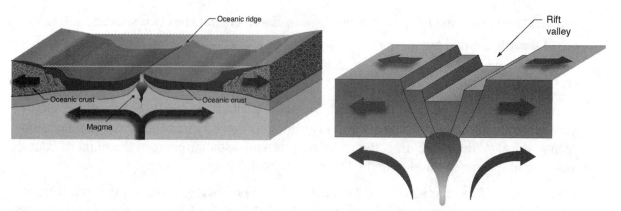

Figure 2.11 Divergent Plate Boundaries. The movement of the asthenospheric rocks away from the top of the rising portion of the mantle convection cell generates stretching forces within the overlying lithosphere that cause it to break into plates, which then move away from each other. Illustrations © Kendall Hunt Publishing Company.

Convergent Plate Boundaries

Convergent boundaries are those where two plates are moving together. When plates come together, one of two things will occur.

(1) One plate will sink beneath the other in a process called subduction (**Figure 2.12**).

(2) If neither plate is light enough to subduct, the plates will collide, generating enormous forces that push mountain ranges into being.

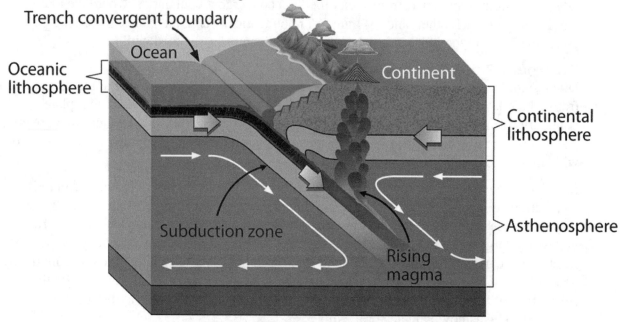

Figure 2.12 Convergent Plate Boundaries. Illustration by Don Vierstra.

Subduction is illustrated in **Figure 2.12**. The deep-sea trenches discovered by early seafloor mapping are the sites of subduction zones and are typical of this type of convergent boundary. Here, oceanic lithosphere has cooled and become dense enough to sink into the asthenosphere.

The forces involved with subduction are tremendous, and plates do not subduct smoothly. As a result of stresses discussed in Lesson 3, plates subduct in periodic jolts, felt at the surface as earthquakes. Earth's largest earthquakes, including the 2011 Japan earthquake, occur at subduction zones. For reasons discussed in the next lesson, most of Earth's volcanoes also occur near subduction zones, including the volcanoes of the "Ring of Fire" encircling the Pacific Ocean.

Transform Plate Boundaries

The third type of plate boundary is the transform boundary, where two tectonic plates slide past each other. The San Andreas Fault that extends from Mexico to Northern California is a transform boundary at the meeting of the Pacific and North American tectonic plates. Here, the Pacific Plate is moving northwest relative to the North American Plate. The plate movement accounts for the large number of earthquakes in California (**Figure 2.13**). The fracture along which the two plates move is called a transform fault.

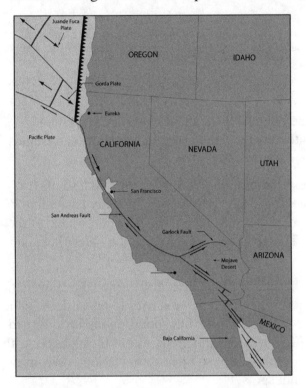
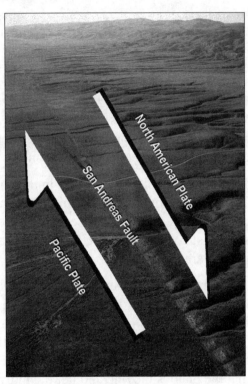

Figure 2.13 Transform Plate Boundary. The best-known fault in North America is the San Andreas Fault, which is a transform plate boundary between the Pacific and North American lithospheric plates. Map (left): From *Planet Earth* by John J. Renton. Copyright © 2002 by John J. Renton. Reprinted by permission of Kendall Hunt Publishing Company; Photo (right): Courtesy of R. E. Wallace/USGS.

Note that the sliding movement between two plates also occurs at the rift zones. This is because the rift is not continuous but rather a series of breaks that are slightly offset from each other. As **Figure 2.14** illustrates, the moving crust causes fractures along lines roughly perpendicular to the rift segments. These fractures are also called transform faults. They allow the jagged edges of the plate to slide past each other as the seafloor spreads. The transform faults at rift zones should not be confused with a transform boundary, even through both describe the same type of sliding movement.

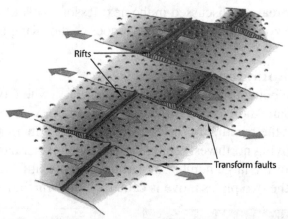

Figure 2.14 Transform Faults. The oceanic rift is not a continuous break but occurs in segments. Fractures in the rift zone called transform faults allow segments of crust to slide past each other as the seafloor spreads. Illustration by John J. Renton.

Hot Spots

While many of Earth's volcanoes occur at plate boundaries, some do not. The Hawaiian Islands are one example of volcanoes in the interior of a plate. Even so, the Hawaiian Islands do provide evidence for plate tectonics.

The Hawaiian Islands are older in the northwest and progressively younger in the southeast. The last and youngest island in the chain is the "Big Island" of Hawaii, the only island that currently has active volcanoes. In addition, there is a new volcano, Lo'ihi, underwater off the southeast coast of the "Big Island" of Hawaii. The volcanoes that created the older Hawaiian islands are all inactive.

A moving plate explains this age difference. The source of the lava that formed the Hawaiian Islands originated from what scientists call a hot spot, where magma rises to the surface in the interior of a tectonic plate. As the Pacific Plate moved over the hot spot, flowing lava formed the individual islands of the Hawaiian chain (**Figure 2.15**). Other chains of volcanoes beneath the surface of the Pacific Ocean have the same northwest to southeast orientation, indicating that the plate is moving and not the individual hot spots.

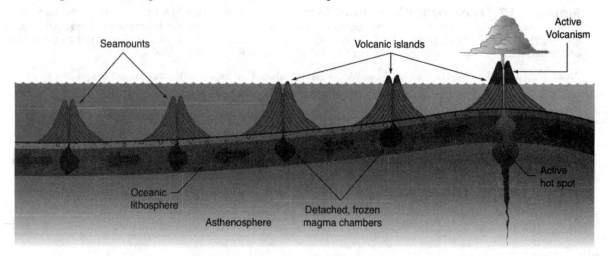

Figure 2.15 Volcanic Arcs. The movement of the Pacific Plate over stationary hot spots produces chains of volcanoes, such as the Hawaiian Islands. Illustration © Kendall Hunt Publishing Company.

Lab Exercise

Lab Exercise: *Plate Boundaries*

Instructions

Step 1:   In this practice step, match the lettered labels in this illustration with the key terms listed below.

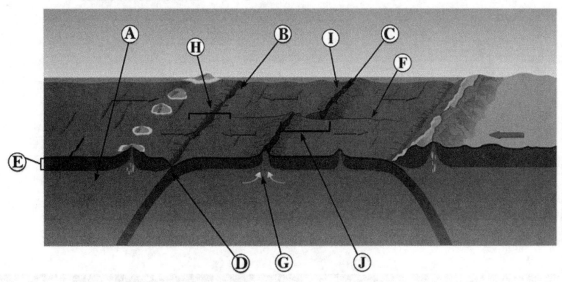

Figure 2.16 Plate Boundaries. Illustration by Don Vierstra.

_____   asthenosphere

_____   lithosphere

_____   transform fault

_____   divergent boundary

_____   convergent boundary

_____   deep-sea trench

_____   subducting plate

_____   rift

_____   oceanic ridge

_____   rising magma

When you've finished this step, check your answers in the student answer key at the end of this lesson or download the answer key from the online supplement.

Step 2:   Make simple illustrations of three types of plate boundaries: convergent, divergent, and transform plate boundaries.

Make sure to include the following:

(1) Arrows indicating the direction in which each plate in the illustration is moving.

(2) Labels, as appropriate, for the following features:

Plate	Deep-sea trench
Asthenosphere	Rift zone
Subducting plate	Upwelling magma

You may make your drawings in the areas indicated below or on a separate piece of paper. Submit your work as directed by your instructor.

Plate Boundary #A—Divergent

Plate Boundary #B—Convergent

Plate Boundary #C—Transform

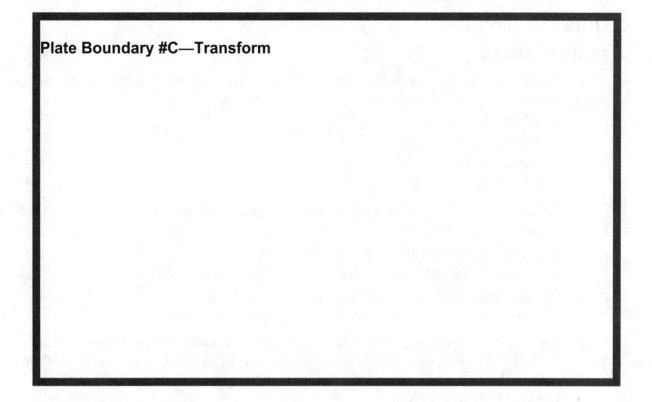

Online Activities

Per your instructor's directions, go to the online supplement for this lab and complete the activities assigned. Viewing the online videos will help you to complete the quiz.

Quiz

Multiple Choice

1. Who proposed that all of the present continents were once joined together in a single supercontinent called Pangaea?
   a. Alfred Wegener
   b. Charles Darwin
   c. Arthur Holmes
   d. Harry Hess

2. The scientific community rejected the theory of plate tectonics because Alfred Wegener could NOT
   a. identify a mechanism to move the continents.
   b. disprove competing theories that were not accepted by scientists.
   c. find geologic similarities on the different continents.
   d. provide evidence that the continents were once joined together.

3. In the 1950s, Harry Hess interpreted the paleomagnetic data of newly formed oceanic crust. He proposed that while new oceanic crust was being formed at the summit of the oceanic ridges, the adjacent oceanic crust was moving laterally away from the ridge. What is this process called?
   a. deep-sea trench
   b. oceanic crust
   c. seafloor spreading
   d. hot spots

4. With the advent of plate tectonics, it was discovered that new oceanic lithosphere is being created at what location?
   a. seafloor spreading
   b. deep-sea trenches
   c. oceanic ridges
   d. continental crust

5. A geologist by the name of Eduard Suess not only noted that the continents fit together but also discovered a fossil that was identical in South America, Africa, and Australia. What type of fossil is it?
   a. palm
   b. seed fern
   c. redwood tree
   d. maple tree

6. In the early 1800s, geologist Eduard Suess suggested that all the southern continents, including Antarctica, had once been joined into one large continent. What name did Suess call the one continent?

   a. Pangaea

   b. Megacontinent

   c. Gondwanaland

   d. Laurasia

7. Alfred Wegener proposed that which two continents "fit together" like a jigsaw puzzle?

   a. Africa and North America

   b. South America and North America

   c. South America and Africa

   d. Eurasia and Africa

8. Wegener proposed that North America and Eurasia had also once been joined into a single large continent called

   a. Megacontinent.

   b. Gondwanaland.

   c. Laurasia.

   d. Rodinia.

9. How many millions of years ago did Wegener propose that Pangaea began to break up and drift across the oceans?

   a. 100 million years

   b. 150 million years

   c. 200 million years

   d. 250 million years

10. In what century did we have new data provided by developing technologies to support the idea that continents could be on the move?

    a. the eighteenth century

    b. the nineteenth century

    c. the twentieth century

    d. the twenty-first century

11. As molten basalt cools, one of the minerals that forms is _____, a black magnetic mineral.

    a. basalt

    b. magnetite

    c. silver

    d. aluminum

12. Which scientist's study of the magnetism of lava flows in Japan resulted in the discovery that Earth's magnetic field appears to have undergone reversals over time— that is, north and south magnetic poles have switched their locations?

    a. Alfred Wegener

    b. Motonori Matuyama

    c. Arthur Holmes

    d. Harry Hess

13. Since the original discovery of magnetic reversals, scientists have shown that these reversals have occurred over the past 10 million years at an average rate of 4 to 5 reversals per million years. The last such reversal occurred about how many years ago?

    a. 680,000 years

    b. 780,000 years

    c. 880,000 years

    d. 980,000 years

14. What is the typical rate of seafloor spreading?

    a. 1 to 5 centimeters per year

    b. 6 to 10 centimeters per year

    c. 11 to 15 centimeters per year

    d. 16 to 20 centimeters per year

15. If new crust is indeed being produced at the rifts, where does the additional older crust go?

    a. It is lost through subduction.

    b. It becomes oceanic crust.

    c. It becomes continental crust.

    d. It is lost through seafloor spreading.

16. Most of Earth's volcanoes are located in the _____ around the Pacific Rim just beyond the oceanic trenches, where magma, which is less dense than the surrounding rock, rises to the surface and erupts.

    a. Ring of Ice

    b. Ring of Water

    c. Ring of Fire

    d. Ring of Steam

17. Stretching forces generated in the lithosphere immediately above the rising portion of a convection cell result in rifting of the lithosphere and ultimate formation of what type of plate boundary?

    a. convergent

    b. transform

    c. divergent

    d. hot spot

18. Pushing forces generated in the lithosphere above the adjacent downgoing portion of the convection cell result in the formation of a deep-sea trench that eventually develops into what type of plate boundary?

    a. convergent
    b. transform
    c. divergent
    d. hot spot

Short Answer
Use Figure 2.17 to answer questions 1 through 4.

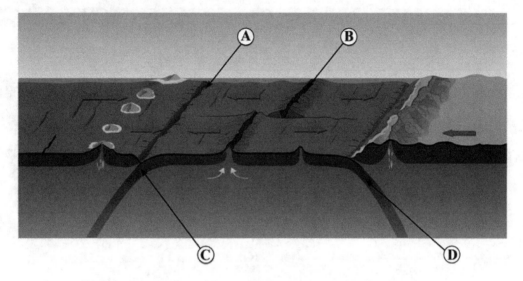

Figure 2.17 Plate Boundaries. Illustration by Don Vierstra.

1. What type of plate boundary is at the point labeled "A"?

2. When an ocean plate converges with another plate, what is created on the seafloor at the line of convergence?

3. What type of plate boundary is being formed at the point labeled "B"?

4. What process is occurring at the points labeled "C" and "D"?

5. How do some rocks store information about Earth's magnetic field?

6. Using an example, describe the geologic process of how volcanic hot spots produce chains of volcanoes.

Student Answer Key

Page 45:

Lab Exercise: *Plate Boundaries*, Step 1

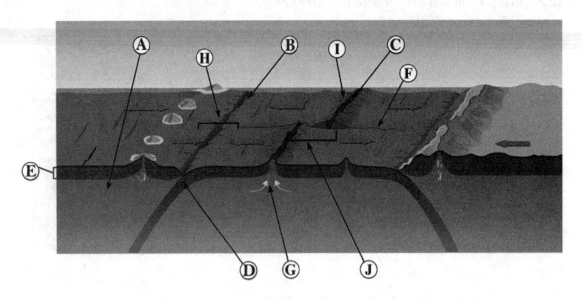

A	asthenosphere
E	lithosphere
F	transform fault
J	divergent boundary
H	convergent boundary
B	deep-sea trench
D	subducting plate
C	rift
I	oceanic ridge
G	rising magma

ROCK DEFORMATION AND MOUNTAIN BUILDING

Lesson 3

AT A GLANCE

Purpose

Learning Objectives

Materials Needed

Overview

    Stress and Strain

    Orientation of Rock Strata

    Types of Structures

    Mountain Building

Lab Exercises

    Lab Exercise #1: *Fault Models*

    Lab Exercise #2: *Strata Folding*

Online Activities

Quiz

    Multiple Choice

    Short Answer

Purpose

This laboratory lesson will familiarize you with the enormous forces that deform the earth's crust and the various geological structures that result from such deformational forces. Understanding how geological forces can change or alter the land surface area will provide insight about how earthquakes are generated, how tectonic movement of plates changes the landscape, and how mountains and hills are created.

Learning Objectives

After completing this laboratory lesson, you will be able to:
- Characterize different types of rock deformation and geologic structures.
- Explain how surface features are caused and created by subsurface movements.
- Demonstrate and analyze fault movement and stress and strain of rock movement.
- Explain mountain building on Earth.

Materials Needed

- ❑ Pencil
- ❑ Eraser
- ❑ Scissors
- ❑ Clear plastic tape
- ❑ Ruler or protractor (in lab kit)
- ❑ Fault model from cardstock paper cut out (in lab kit)
- ❑ Three different colors of flexible foam (in lab kit)

Overview

While it may seem as if rocks are timeless and unchanging, in fact, rocks in Earth's lithosphere undergo almost constant change. Some of the changes occur at the surface as the result of water and wind. Many other changes occur as the result of plate tectonics, forming the variation in Earth's landscape. Rocks that have undergone a change in their original size or shape are said to be strained. The change in size and shape is generally caused by the enormous forces created during movements of the earth's tectonic plates over long periods of geological time. Such slowly applied but immense forces can build huge mountain ranges or carry rock deep underground, permanently deforming the rock.

Geologists seek to unravel the history of rock deformation by studying the causes and effects of forces placed upon rocks composing Earth's lithosphere. The force responsible for such changes is called stress; strain is the response of rocks to stress.

Stress and Strain

Simply stated, stress is a directed force, externally applied, that tends to change the size and/or shape of an object; in this case, rocks. The type of strain the rocks exhibit depends on the type of stress applied. There are three types of stress: tension, compression, and shear. The difference between them is the direction of the forces being applied.

Tension is a pulling apart or stretching force that can deform an object by making it longer and thinner (**Figure 3.1**). This is the same force involved in pulling apart dough or stretching a rubber band. It is the type of stress associated with divergent plate boundaries where plates are moving away from each other and with areas where rifting is occurring, such as the East Africa rift zone discussed in Lesson 2.

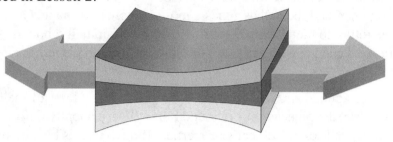

Figure 3.1 Tension. The forces involved in tension operate directly opposite and away from each other. Anytime you pull on an object, you are applying tensional forces. Illustration by Don Vierstra.

In compression, the forces act toward and directly opposite each other. Compression shortens and thickens an object (**Figure 3.2**). It is often called a pushing force. A trash compactor applies compression by pushing in on all sides, pushing trash into a compact block. Compression is the stress associated with convergent plate boundaries, where plates are moving toward each other. Earth's most prominent mountain ranges are the result of compression, as discussed later in this lesson.

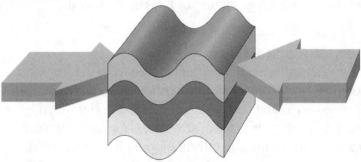

Figure 3.2 Compression. The forces involved in compression act toward and directly opposite each other. Illustration by Don Vierstra.

Shear stress is a tearing force in which one side of an object is pushed or pulled one way while the other side is pushed or pulled in the opposite direction (**Figure 3.3**). Shear is the force that tears paper and gives scissors their power to cut through things. In fact, scissors are often called shears after the type of stress they use to accomplish their mission. Shear is the stress associated with transform plate boundaries where two plates are sliding past each other.

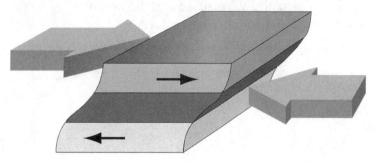

Figure 3.3 Shear Stress. The forces in shear also are directed toward each other, but are not directly opposed. Illustration by Don Vierstra.

Lesson 3/Rock Deformation and Mountain Building

Any stress, no matter how small, produces some strain. Strain is the response to stress; it is also referred to as deformation. The type of strain an object undergoes depends on its properties. Most objects do not permanently deform immediately; they undergo a temporary deformation when stress is initially applied and then go back to their original shape when the stress is released. For example, if a foam ball is squeezed in your hand, the particles in the ball are squeezed together, and the ball assumes a more compressed shape. Once the stress is removed, the particles return to their original spacing, and the foam ball rebounds to its original size. This type of strain is called elastic strain because the deformation is temporary and the object recovers fully.

However, if the applied stress is great enough, it may exceed what is called the object's elastic limit; in this case, the object won't rebound but will permanently deform. The type of deformation depends on the object's properties. If the object is brittle, like glass or cold taffy, it will permanently deform by breaking or fracturing when stressed; this is called brittle strain.

If an object is ductile, like modeling clay or warm taffy, it will permanently change its shape in response to stress; this is called plastic strain. A ductile object subjected to compression will change its shape, like the foam ball, or fold. If subjected to tension, a ductile object, like warm taffy, will permanently stretch into a longer, thinner shape. Shear stress will fold or buckle a ductile material.

Many of Earth's landforms are the visible results of plastic and brittle strain, the products of rock strata, or layers, breaking, folding or buckling under the stresses produced by plate tectonics.

Orientation of Rock Strata

Strained rock strata can take on any orientation during deformation from slightly tilted to vertical to completely upside down. Furthermore, because stress can vary over distance, the resulting deformed rock strata can change orientation significantly in a relatively short distance. Part of the work of geologists is to identify, describe and map such rock formations. Their descriptions and drawings include information on how the rock strata are oriented relative to compass directions and relative to a horizontal plane.

Strike and Dip

A plane is a flat, two-dimensional surface, like the surface of a piece of paper. A horizontal plane is one that is oriented parallel to Earth's surface. Many geologic structures have planar components; the surfaces of different rock strata are planar as are the surfaces of rock fractures.

Strike and dip are measurements that allow the spatial orientation of geological features to be described in such a manner that anyone can visualize the orientation without actually seeing the feature. It's easiest to visualize what strike and dip are by using a common object with both strike and dip—the roof of a house (**Figure 3.4**).

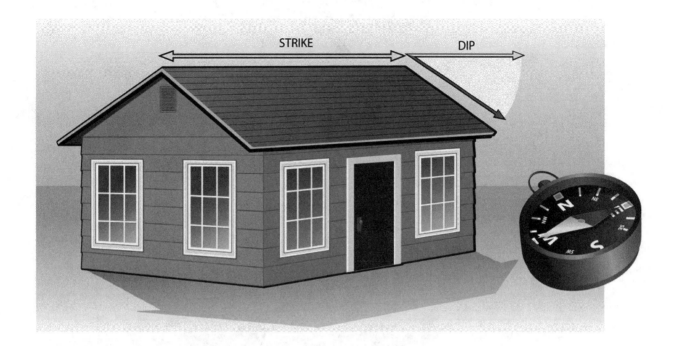

Figure 3.4 Strike and Dip of a Roof. Roofs exhibit both strike and dip. The strike is the orientation of the peak of the roof. The dip is the angle of the roof's slope. Illustration by James H. Brown and Don Vierstra.

Strike is the orientation of a structure relative to truth north; in terms of our example, the direction in which the peak of the house points. More specifically, strike is the compass orientation of the line of intersection between the plane (or surface) of the structure and a horizontal plane.

Dip is the angle between the plane of the structure (the roof's surface) and a horizontal plane. It is expressed as degrees. However, because a plane can dip away from a line in two possible directions, dip measurements often include a direction.

Strike and dip allow a geologist to describe a planar surface in three-dimensional space (see **Figure 3.5**). Note that the directions of strike and dip are always perpendicular to each other.

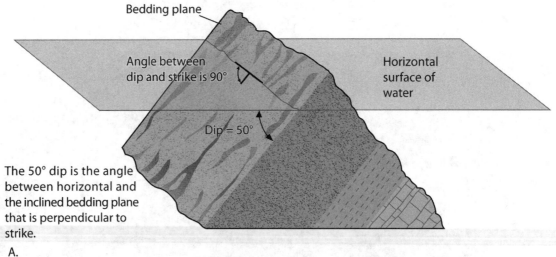

Bedding plane

Angle between dip and strike is 90°

Horizontal surface of water

Dip = 50°

The 50° dip is the angle between horizontal and the inclined bedding plane that is perpendicular to strike.

A.

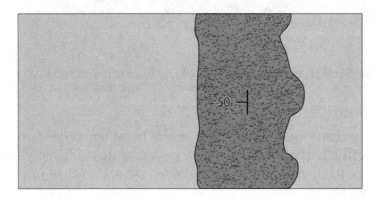

50°

B.

Figure 3.5 Strike and Dip of a Geological Structure. The two measurements of strike and dip allow a geologist to describe a planar surface in three-dimensional space. Note that the directions of strike and dip are always perpendicular to each other. Illustration by Don Vierstra.

Another good way to envision strike and dip is by means of a book and a table, using the table top as the imaginary horizontal plane. To do this, designate the side of the table nearest you as south and the opposite side as north, the right side as east and the left side as west. Place the book vertically on its edge on the table so that it is aligned in a north-south direction. The book now has a north-south strike and a dip of 90 degrees. Next, tilt the book about 45 degrees to the right (to the east). Notice that that the book is inclined downward toward the left (to the west). The book still has a north-south strike, but the dip angle is now 45 degrees and the dip direction is to the west.

Types of Structures

Rocks deform under stress. The type of deformation depends on the properties of the rock and the type of stress. The three most common structures formed through rock deformation are folds, faults, and joints.

Folds

Folds are bends in rock strata and generally represent the ductile strain of the rocks in response to compression within Earth's crust. Folds occur at all scales and range in size from small warps visible within individual outcrops to large-scale features that extend over many miles. Rock strata that fold upward form arch-like structures, called anticlines, with limbs, or sloping sides of the fold, extending downward from the center of the arch (the axis) (**Figure 3.6**). Rock strata that fold downward form trough like structures, called synclines, with limbs that extend upward from the axis. With an anticline, the older rock strata appear on the inside of the fold and the younger strata are on the outside of the fold. The opposite occurs with a syncline; the younger strata appear inside the fold and the older strata are on the outside of the fold.

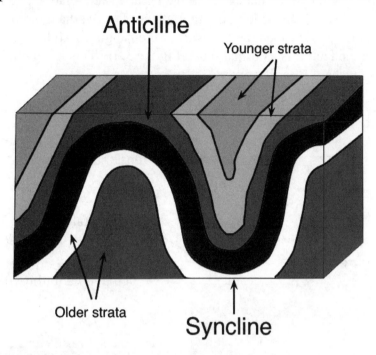

Figure 3.6 Anticline and Syncline. The most commonly observed folds are anticlines and synclines. Note that younger rock strata appear on the inside of the syncline whereas older rock strata appear on the inside of the anticline. Illustration: Courtesy of USGS.

Folds take many forms. Geologists describe a fold in terms of how its axial plane and limbs are positioned. An axial plane is an imaginary plane that runs through the center of the fold (**Figure 3.7**). If the axial plane divides the fold into two essentially equal and identical halves, the fold is said to be symmetric. If the two halves are different in area and/or shape, the fold is asymmetric. In general, symmetric folds form as the result of compression, whereas asymmetric folds form as the result of shear.

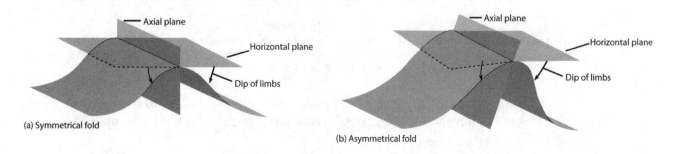

Figure 3.7 Types of Folds. Folds are described as being symmetric and asymmetric, depending on the relationship of the axial plane and the limbs relative to a horizontal plane. Illustration © Kendall Hunt Publishing Company.

Lesson 3/Rock Deformation and Mountain Building   **59**

Faults

Rocks that respond to stress with brittle strain may form a fault. A fault is a break or rupture in the earth's crust, along which there has been movement. Therefore, faulting displaces the rock on one side of the break, or fault plane, relative to the rock on the other side.

There are three types of faults: (1) normal faults, (2) reverse/thrust faults, and (3) strike-slip faults. Each is named by observing the movement between the two sides of the fault. Normal faults and reverse/thrust faults have fault planes at a slant relative to a horizontal plane. This forms two blocks on either side of the fault plane. The overhanging block is called the hanging wall; the other block is called the footwall. If the hanging wall has moved downward relative to the footwall, the fault it is a normal fault (**Figure 3.8**) . This is generally caused by tension in the shallow reaches of Earth's crust.

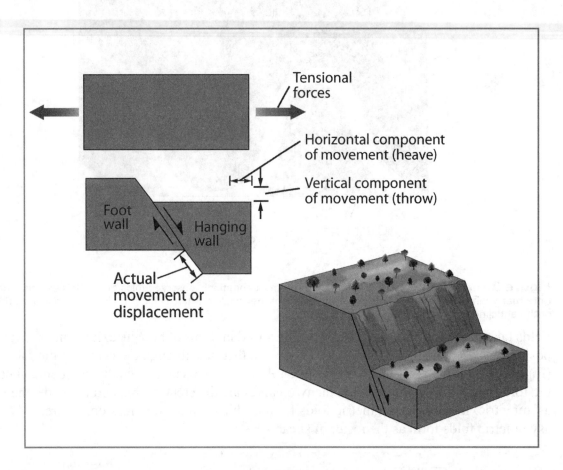

Figure 3.8 Normal Fault. Tensional stresses produce normal faults in which the hanging wall moves down relative to the footwall. From *Planet Earth* by John J. Renton. Copyright © 2002 by John J. Renton. Reprinted by permission of Kendall Hunt Publishing Company.

On the other hand, if the hanging wall has moved upward relative to the footwall, then the fault is a reverse or thrust fault (**Figure 3.9**), which is generally caused by compression in the upper reaches of Earth's crust.

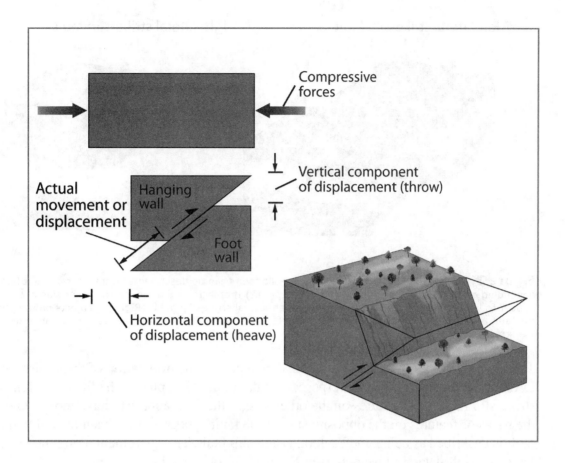

Figure 3.9 Reverse or Thrust Fault. Reverse or thrust faults form by the brittle failure of rocks under compression. As a result of the movement, the length of Earth's crust is shortened and the crust is locally thickened. From *Planet Earth* by John J. Renton. Copyright © 2002 by John J. Renton. Reprinted by permission of Kendall Hunt Publishing Company.

The difference between a reverse and thrust fault is the dip of the fault plane. If the dip is less than 45°, the fault is a thrust fault; if the angle is greater than 45°, the fault is a reverse fault. In both thrust and reverse faults, the hanging wall has moved up relative to the footwall. Faults associated with zones of subduction are either thrust or reverse faults, depending on the dip. If a thrust fault does not intersect with Earth's surface, the fault is referred to as a blind thrust fault. Blind thrust faults are common in California, and they are often discovered only after they produce an earthquake.

Strike-slip faults are the result of shear forces (**Figure 3.10**). Because the fault plane is vertical or near-vertical and the displacement is horizontal (sideways), there is no hanging wall or footwall. A transform boundary, discussed in Lesson 2, is an example of a strike-slip fault.

left-lateral strike-slip fault   right-lateral strike-slip fault

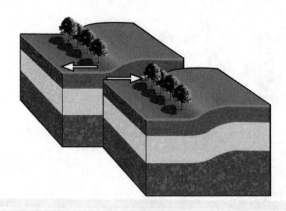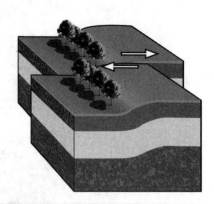

Figure 3.10 Strike-Slip Faults. Strike-slip faults can be described as right lateral or left lateral depending on the direction of relative movement. (Left) If objects, like trees, on the other side of the fault appear to have moved to the left, the fault is a left lateral strike-slip fault. (Right) If objects on the other side of the fault appear to have moved to the right, the fault is a right lateral strike-slip fault. Illustration by Don Vierstra.

There are two types of strike-slip faults, depending on the orientation of the shear forces. Looking directly across the fault (perpendicular to the fault plane), the fault is a right-lateral strike-slip fault if the features on the other side of the fault appear to have moved to the right. If the rocks or features on the opposite side of the fault appear to have been moved to the left, the fault is described as being a left-lateral strike-slip fault. The movement along the San Andreas Fault is a right-lateral movement (see **Figure 2.13** on page 43).

Joints

Joints are fractures, a form a brittle strain, along which there has been no movement or displacement. Joints are the most common of all geologic structures and form under a wide variety of conditions, all of which involve stress. It is the rare rock bed that has no cracks or other fractures!

Most joints form as the result of compression. Compression has the effect of shortening the rock parallel to the stress and lengthening it perpendicular to the stress. Shear forces also come into play as the rock is being pushed and squeezed. The combined force of these stresses tends to fracture the rocks in a perpendicular pattern (**Figure 3.11**). Cooling rock may also crack, forming joints in the new rock body.

Figure 3.11 Joints. Joints, like these in California's Joshua Tree National Park, are fractures in rock where there has been no movement or displacement. Joints are often perpendicular to each other. Photo: Courtesy of Susan Wilcox.

Another type of joint occurs when lava cools so quickly that it cracks. In this case, the joints, called columnar jointing, take the form of long hexagonal columns. California's Devil's Postpile and Wyoming's Devil's Tower are both structures composed of columnar joints.

Mountain Building

The same forces that produce folds and faults on a local scale have created mountains on a regional scale. These forces often occur at plate boundaries. Earth's longest mountain range, the oceanic ridge, is formed at a divergent boundary from the extrusion of lava that forms new seafloor. Although magma is produced by divergent plate boundaries, and mountains, like the oceanic ridge, can form at divergent plates, convergent plate boundaries are responsible for most of Earth's mountain building. Certain types of mountains are associated with different plate combinations that occur at the convergent boundaries.

Ocean-Ocean Mountain Building

Ocean-ocean mountain building describes how mountains form at a convergent boundary where the leading edge of both plates is oceanic lithosphere. In this case, one tectonic plate subducts beneath the other plate, creating pressures and temperatures great enough to melt some of the rocks into magma. Magma is less dense than the surrounding rock, so it moves toward the surface; when it breaks through the surface, the result is a volcano (**Figure 3.12**). At ocean-ocean convergent boundaries, the volcanoes build beneath the surface of the ocean. Some will become large enough to break through the surface as volcanic islands. A string of volcanic islands, called a volcanic arc, is typical of ocean-ocean mountain building. The Philippine Islands, Japan, and Alaska's Aleutian Islands are all examples of volcanic arcs formed over subduction zones at ocean-ocean convergent plate boundaries.

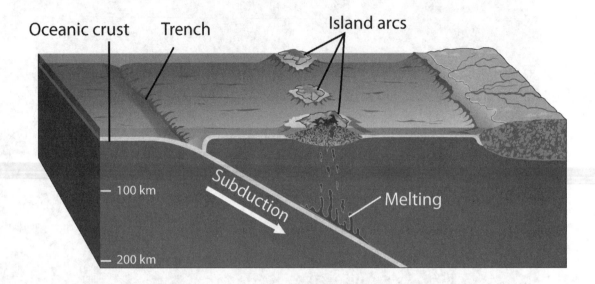

Figure 3.12 Ocean-Ocean Mountain Building. Volcanic island arcs form when magma near a subduction zone rises and erupts. The magma accumulates until it breaks through the surface, forming a volcanic island. A string of such volcanic islands, the island arc, is typical of ocean-ocean convergence boundaries. Illustration by Don Vierstra.

Ocean-Continent Mountain Building

Ocean-continent mountain building describes what occurs at a convergent boundary when the leading edge of one plate is oceanic lithosphere, and the leading edge of the other plate is continental lithosphere. Because continental lithosphere is less dense than oceanic lithosphere, the latter will subduct beneath the former. As with ocean-ocean mountain building, subduction creates pressures and temperatures great enough to melt some rock into magma. When the magma reaches the surface, however, it extrudes through continental crust, and the lava forms a continental volcanic arc (see **Figure 3.13 B**). The volcanoes develop into a mountain range along the continental margin with folded and faulted rock strata located inland. Examples of mountains created by the ocean-continent orogenic style are the Andes Mountains of South America, the Cascade Mountains of Northern California, Oregon, and Washington State, and Alaska's Aleutian Range.

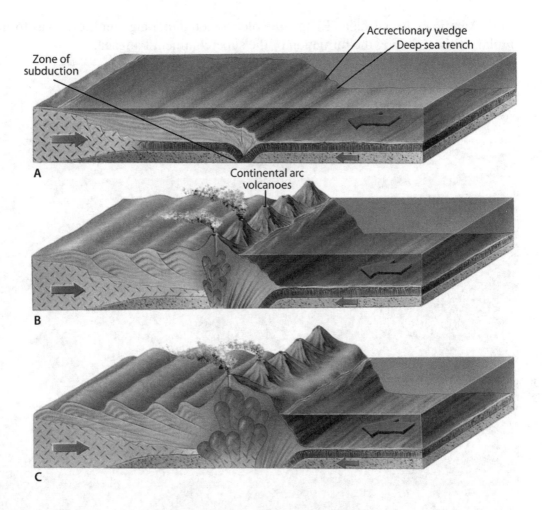

Accretionary wedge
Deep-sea trench
Zone of
subduction

A

Continental arc
volcanoes

B

C

Figure 3.13 Ocean-Continent Mountain Building. Ocean-continent mountain building begins when oceanic lithosphere subducts beneath continental lithosphere (A). The subduction results in rock melting to become magma, which rises to the surface and erupts (B). Over time, solidified lava from the eruptions accumulates to form a range of volcanic mountains called a continental arc. From *Planet Earth* by John J. Renton. Copyright © 2002 by John J. Renton. Reprinted by permission of Kendall Hunt Publishing Company.

Continent-Continent Mountain Building

Continent-continent mountain building occurs at convergent boundaries where the leading edges of both lithospheric plates are continental lithosphere. In this case, neither plate has enough density to subduct, so the two continents collide. The collision of the two plates involves tremendous compressive forces that push the rocks of both plates downward and upward, forming Earth's most impressive mountain ranges **(Figure 3.14)**.

This type of mountain building by compressional forces acting upon layers of rock at plate boundaries is called orogeny. The word *orogeny* comes from the Greek word *oros* for mountain and the word *genesis*, which means birth or origin, so the process of mountain building comes from forces or events that produce severe structural deformation of Earth's crust by tectonic plate movement.

Mountains formed from compression at continent-continent convergent boundaries are called foldbelt mountains. The Himalayan Range, perhaps Earth's most dramatic mountain landscape, is currently being formed by the collision of the India plate with the Eurasian plate.

North America's Appalachian Range, the oldest mountain range on Earth, was formed by a similar collision during the formation of the supercontinent Pangaea.

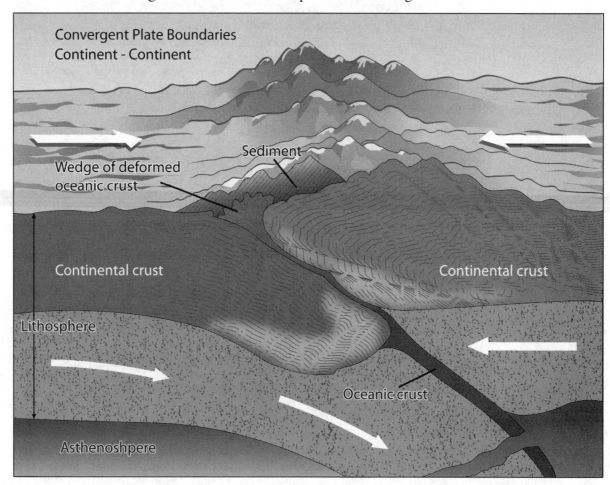

Figure 3.14 Continent-Continent Orogeny. Foldbelt mountains form when continents collide and compression forces the crust both upward and downward into folds. Illustration by Don Vierstra.

Lab Exercises

Lab Exercise #1: *Fault Models*

In this lab exercise, you will demonstrate the movement and effects of three different types of faults. Each fault will show some form of surface feature that provides clues to determining the type of fault it is. Follow the steps below and record your observations in the space provided or on a separate piece of paper.

Reference **Figures 3.8**, **3.9**, and **3.10** (i.e., three different types of faults: normal, reverse, and strike-slip) as a guide to assist you in answering the questions pertaining to this exercise. *Note: The scale of the model is 1 millimeter (mm) = 1 meter (m).*

Make sure to save your results and observations. You will use the data to answer the questions at the end of this lesson.

Instructions and Observations

Step 1: Locate the fault model image (on white cardstock) found in your lab kit.

Step 2: Using a pair of scissors, cut along the outside borders of the image. Then, cut through the image along the dotted line. You will now have two pieces of the fault model.

Step 3: Fold each half of the model, so that the rock layers match. You should fold down three sides on one piece of the fault model and proceed to fold the three sides on the other piece of the fault model.

Step 4: Use transparent tape to connect the corners on each piece.

Step 5: Place the two pieces on a flat surface so that they stand up and are arranged side by side. Make sure that the rock layers match.

Step 6: Move the two pieces so that point S is next to point T. Use a ruler or protractor to determine how far the rock mass would move along a lateral fault if point S moved next to point T. Record the measurement.
Measured distance (on the model) = _____ mm

Step 7: Convert the measured distance to actual distance. (Remember the scale is 1 mm on the model equals 1 m of actual distance.)

Actual distance = _____ m

Step 8: Move the two pieces back to their original position.

Step 9: Move the two pieces so that point S is next to point R. Use a ruler or protractor to determine how far the rock mass would move along a lateral fault if point S moved from the initial position to point R. Record the measurement.

Measured distance (on the model) = _____ mm

Step 10: Convert the measured distance to actual distance. (Remember the scale is 1 mm on the model equals 1 m of actual distance.)

Actual distance = _____ m

Step 11: Observe the model. If point S moved to point R suddenly because of slippage along the fault, and if the stream flowed to the west, where would the stream flow relative to the landmarks in the area (railroad, power lines, tree, road, etc.)?

Step 12:   Observe the model.

1. Which type of fault would exist if point S moved near point R?

    a. a normal fault

    b. a reverse fault

    c. a right-lateral strike-slip fault

    d. a left-lateral strike-slip fault

2. What type of plate boundary would be related to this type of fault?

    transform plate boundary

Step 13:   Hold the model in your hands, with one piece in each hand. Starting from the original position of the two pieces, slide one side down and the other side up so that point E and point F are aligned horizontally.

1. Which type of fault would exist if point E moved near point F?

    a. a thrust fault

    b. a normal fault

    c. a right-lateral strike-slip fault

    d. a left-lateral strike-slip fault

2. What type of plate boundary would be related to this type of fault?

    divergent plate boundary

Step 14:   Starting from the original position of the two pieces, move one side down and the other side up so that point G and point H are aligned horizontally.

1. Which type of fault would produce movement along the fault to place point G near point H?

    a. a reverse fault

    b. a normal fault

    c. a right-lateral strike-slip fault

    d. a left-lateral strike-slip fault

2. What type of plate boundary would be related to this type of fault?

    convergent plate boundary

Lab Exercise #2: *Strata Folding*

Some mountains are formed by folding and others are formed by fault movement. Deep within Earth's crust the layers of rock become extremely hot, thus becoming very pliable. Many mountain ranges are formed when the colliding crust is compressed and pushed upward and downward.

In this lab exercise, you will be creating and demonstrating two different kinds of folds. Each fold will demonstrate some form of surface feature that provides you a clue for identifying what type of fold is being expressed.

Make sure to save your observations. You will use the data to answer the questions at the end of this lesson.

Instructions and Observations

Step 1: Locate the three pieces of different colored flexible foam found in your lab kit.

Step 2: Stack the three pieces, one on top of the other. (Color order does not matter.) The different colored foam layers represent rock layers with the youngest rock layer on the top and the oldest rock layer on the bottom.

Step 3: Hold the left side of the foam stack with your left hand and hold the right side of the foam stack with your right hand.

Step 4: Hold up the layers of foam in the air with both hands. Move both hands toward each other (move your left hand toward the right and your right hand toward the left) to create an upward bulge in the middle.

Step 5: Observe the fold from the side.

1. What type of fold is created from the deformation of the rock layers?
   a. syncline
   b. anticline
   c. asymmetric fold
   d. recumbent fold

2. Where on or within the fold are the youngest rocks located in this type of fold?
   a. They appear on the outside of the fold.
   b. They appear on the inside of the fold.
   c. Neither of the above because the youngest rocks are in the middle layer.
   d. The rocks are evenly distributed throughout the fold by age.

Step 6: Now hold up the layers of foam in the air with both hands. Move both hands toward each other (move your left hand toward the right and your right hand toward the left) so as to create a downward bulge in the middle.

Step 7: Observe the fold from the side.

1. What type of fold is created from the deformation of the rock layers?
   a. syncline
   b. anticline
   c. asymmetric fold
   d. recumbent fold

2. Where on or within the fold are the youngest rocks located in this type of fold?
   a. They appear on the outside of the fold.
   b. They appear on the inside of the fold.
   c. Neither of the above because the youngest rocks are in the middle layer.
   d. The rocks are evenly distributed throughout the fold by age.

Step 8: Answer the following questions about types of stress.

1. What type of stress did you apply to the rock layers in Steps 4 and 6?

    a. tension

    b. shear

    c. sporadic

    d. compression

2. What type of mountains are formed when the type of stress applied in Steps 4 to 6 occurs during a continent-continent collision?

    a. volcanic island arcs

    b. continental arcs

    c. foldbelt mountains

    d. rocky mountains

Online Activities

Per your instructor's directions, go to the online supplement for this lab and complete the activities assigned. Viewing the online videos will help you to complete the quiz.

Quiz

Multiple Choice

Questions 1 through 9 are based on **Lab Exercise #1:** *Fault Models.*

1. What is your answer from **Lab Exercise #1, Step 6**?

    a. 20 mm

    b. 25 mm

    c. 28 mm

    d. 30 mm

2. What is your answer from **Lab Exercise #1, Step 7**?

    a. 20 m

    b. 22 m

    c. 25 m

    d. 50 m

3. What is your answer from **Lab Exercise #1, Step 9**?

    a. 20 mm

    b. 25 mm

    c. 28 mm

    d. 30 mm

4. What is your answer from **Lab Exercise #1, Step 10**?

    a. 10 m

    b. 20 m

    c. 28 m

    d. 30 m

5. What is your answer from **Lab Exercise #1, Step 11**?

    a. The stream would flow into the railroad tracks.

    b. The stream would flow into the road.

    c. The stream would flow under the power lines.

    d. The stream's course would not change.

6. Record your answer from **Lab Exercise #1, Step 12, Question 1**.Which type of fault would exist if point S moved near point R?

    a. a thrust fault

    b. a normal fault

    c. a left-lateral strike-slip fault

    d. a right-lateral strike-slip fault

7. Record your answer from **Lab Exercise #1, Step 12, Question 2**.What type of plate boundary would be related to this type of fault?

    a. divergent boundary

    b. convergent boundary

    c. transform boundary

    d. continental boundary

8. Record your answer from **Lab Exercise #1, Step 13, Question 1**.Which type of fault would exist if point E moved near point F?

    a. a thrust fault

    b. a normal fault

    c. a right-lateral strike-slip fault

    d. a left-lateral strike-slip fault

9. Record your answer from **Lab Exercise #1, Step 13, Question 2**. What type plate boundary would be related to this type of fault?

    a. divergent boundary

    b. convergent boundary

    c. transform boundary

    d. continental boundary

10. Record your answer from **Lab Exercise #1, Step 14, Question 1**. Which type of fault would produce movement along the fault to place point G near point H?

    a. a reverse fault

    b. a normal fault

    c. a right-lateral strike-slip fault

    d. a left-lateral strike-slip fault

11. Record your answer from **Lab Exercise #1, Step 14, Question 2**. What type plate boundary would be related to this type of fault?

    a. divergent boundary

    b. convergent boundary

    c. transform boundary

    d. continental boundary

For Questions 12 through 17, fill in the chart below with the stresses that cause movements along a fault.

Type of Fault	Cause of Fault Motion Stress: Which type of stress causes the faults to move? (Shear, Compression, or Tension)	Use arrows to describe the direction of the stress of the fault. Examples of rock stress: Tension ← → Compression → ← Shear →
Normal Fault	12.	15.
Thrust / Reverse Fault	13.	16.
Left-Lateral or Right-Lateral Strike-Slip Fault	14.	17.

12. Normal Fault

    a. shear

    b. compression

    c. tension

13. Thrust/Reverse Fault

    a. shear

    b. compression

    c. tension

14. Normal Fault

    a.

    b. ← →|

    c. →

    d. ⇄←

15. Thrust/Reverse Fault

    a. $\longrightarrow\!\!\longleftarrow$ |

    b. $\longleftarrow\quad\longrightarrow$ |

    c. $\overset{\longleftarrow}{\longrightarrow}$ |

    d. $\overset{\longrightarrow}{\longleftarrow}$ |

16. Left-Lateral Strike-Slip Fault

    a. $\longrightarrow\!\!\longleftarrow$ |

    b. $\longleftarrow\quad\longrightarrow$ |

    c. $\overset{\longleftarrow}{\longrightarrow}$ |

    d. $\overset{\longrightarrow}{\longleftarrow}$ |

17. Right-Lateral Strike-Slip Fault

    a. $\longrightarrow\!\!\longleftarrow$ |

    b. $\longleftarrow\quad\longrightarrow$ |

    c. $\overset{\longleftarrow}{\longrightarrow}$ |

    d. $\overset{\longrightarrow}{\longleftarrow}$ |

Questions 18 through 23 are based on **Lab Exercise #2:** *Strata Folding.*

18. Record your answer from **Lab Exercise #2, Step 5, Question 1**. What type of fold is created from the deformation of the rock layers?

    a. syncline
    b. anticline
    c. asymmetric fold
    d. recumbent fold

19. Record your answer from **Lab Exercise #2, Step 5, Question 2**. Where on or within the fold are the youngest rocks located in this type of fold?

    a. They appear on the outside of the fold.

    b. They appear on the inside of the fold.

    c. Neither of the above because the youngest rocks are in the middle layer.

    d. The rocks are evenly distributed throughout the fold by age.

20. Record your answer from **Lab Exercise #2, Step 7, Question 1**. What type of fold is created from the deformation of the rock layers?

    a. syncline

    b. anticline

    c. asymmetric fold

    d. recumbent fold

21. Record your answer from **Lab Exercise #2, Step 7, Question 2**. Where on or within the fold are the youngest rocks located in this type of fold?

    a. They appear on the outside of the fold.

    b. They appear on the inside of the fold.

    c. Neither of the above because the youngest rocks are in the middle layer.

    d. The rocks are evenly distributed throughout the fold by age.

22. Record your answer from **Lab Exercise #2, Step 8, Question 1**. What type of stress did you apply to the rock layers in Steps 4 and 6?

    a. tension

    b. shear

    c. sporadic

    d. compression

23. Record your answer from **Lab Exercise #2, Step 8, Question 2**. What type of mountains are formed when the type of stress applied in Steps 4 and 6 occurs during a continent-continent collision?

    a. volcanic island arcs

    b. continental arcs

    c. foldbelt mountains

    d. rocky mountains

24. The best description of the relative movement across the normal fault depicted in **Figure 3.6** is that the

    a. hanging wall has moved down relative to the footwall.

    b. footwall has moved down relative to the hanging wall.

    c. hanging wall and footwall have moved horizontally in a right-lateral sense.

    d. hanging wall and footwall have moved horizontally in a left-lateral sense.

25. The San Andreas Fault is a
    a. thrust fault.
    b. right-lateral strike-slip fault.
    c. left-lateral strike-slip fault.
    d. normal fault.

26. Faults associated with zones of subduction are _____ faults depending on the angle of the dip.
    a. normal
    b. reverse or thrust
    c. left-lateral strike-slip
    d. right-lateral strike-slip

27. _____ are bends in the rock strata and generally represent the ductile response of rock strata to compression within the earth's crust.
    a. Joints
    b. Normal faults
    c. Folds
    d. Reverse faults

28. A continent-continent collision produces a
    a. trench.
    b. spreading center.
    c. mountain range.
    d. volcanic island arc.

29. The Andes Mountains of South America and the Cascade Mountains of Northern California, Oregon, and Washington State are produced by what type of plate tectonic occurrence?
    a. ocean-ocean convergence
    b. ocean-continent convergence
    c. continent-continent collision
    d. divergent plate boundary

30. When a denser oceanic plate moves underneath a less-dense continental plate, what two types of geologic structures are produced during this plate movement?
    a. deep-sea trench and continental arc
    b. deep-sea trench and island arc
    c. valley and hill
    d. hotspot and rift valley

31. The Appalachian Mountains were uplifted by what type of mountain building scenario?
    a. ocean-ocean plate convergence
    b. continent-continent plate collision
    c. ocean-continent plate convergence
    d. normal faulting

32. Whether it is steel and aluminum, cookie dough, cheese, or paper, few substances stay the same when under stress; the response of a substance to stress is called

    a. tension.
    b. compression.
    c. shear.
    d. strain.

33. Temperature often affects whether a substance will break or ooze. Taffy is an example of a substance that is brittle when cold but _____ at room temperature.

    a. strain
    b. ductile
    c. brittle
    d. strong

34. Rock has a response to stress similar to that of taffy. Crust that is close to the surface and cooler is more likely to be brittle and to break when stressed. Increase the pressure, however, and the rock will heat up, becoming ductile. Under heat and pressure, it will change its shape rather than break. Both brittle and ductile responses to stress are evident in

    a. jointing.
    b. mountain ranges.
    c. seafloor spreading.
    d. stream flow.

35. The orientation of a line of intersection between a horizontal plane and a tilted rock bed relative to true north is called the bed's

    a. strike.
    b. tilt.
    c. dip.
    d. perspective.

36. The angle expressing the amount of incline between a tilted rock bed and a horizontal plane is called the bed's

    a. strike.
    b. tilt.
    c. dip.
    d. perspective.

Short Answer

1. What is the difference between stress and strain?

2. Explain what is meant by an elastic limit.

3. A fault is an example of what kind of strain? Of what kind of strain is a fold an example?

4. How does a joint differ from a fault?

5. Why are faults and folding generally associated with mountain ranges?

6. In what mountain-building scenario does orogeny occur?

7. What type of mountain building produces foldbelt mountains?

8. Where in North America would you find foldbelt mountains?

9. What kind of mountain building produces a volcanic arc?

10. Where in North America would you find a volcanic island arc?

11. Where in North America would you find a continental arc?

EARTHQUAKES AND SEISMOLOGY

Lesson 4

AT A GLANCE

Purpose

Learning Objectives

Materials Needed

Overview

    Seismic Waves

    Reading a Seismogram

    Locating the Epicenter

    Wadati-Benioff Zone

Lab Exercises

    Lab Exercise #1: *Time Lag*

    Lab Exercise #2: *Locating the Epicenter of an Earthquake*

    Lab Exercise #3: *USGS Earthquake Hazards Program*

Online Activities

Quiz

    Multiple Choice

    Short Answer

Purpose

This laboratory lesson will familiarize you with geologic concepts pertaining to earthquakes and seismology.

Lesson Objectives

After completing this laboratory lesson, you will be able to:

- Identify various kinds of seismic waves.
- Interpret the results from a seismograph and locate the epicenter of an earthquake.
- Explain how the locations of most earthquakes provide support for plate tectonic theory.

Materials Needed

- ❏ Drafting compass (in lab kit)
- ❏ Ruler (in lab kit)
- ❏ Pencil and eraser
- ❏ Calculator

Overview

Earthquakes are ground vibrations caused by the release of energy from fault movements, asteroid impacts, volcanic eruptions, explosions, and movements of magma. This laboratory lesson focuses specifically on earthquakes produced from fault movements.

Earthquakes occur as the result of built-up stress. As discussed in Lesson 2, brittle rocks subjected to stress will undergo elastic deformation during which they store the energy of the stress until they reach their elastic limit. Once the stress exceeds the elastic limit, the stored energy is released. If the built-up stress is along a fault, some of the released energy will cause the rocks to slip along the fault plane, rupturing the fault. If there is no fault, the released energy will fracture the rocks to create one. The ruptures may be localized or travel along a long segment of a fault (**Figure 4.1**). Either way, the remaining energy is released as seismic waves that we feel as ground movement: an earthquake.

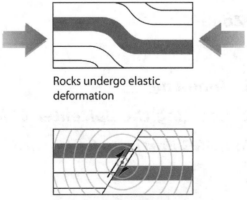

Rocks undergo elastic deformation

Rocks break and move

Figure 4.1 Fault Rupture. Rocks under stress may deform (top), but if they are brittle, they can only deform until they've reached their elastic limit. Once the elastic limit is exceeded (bottom), the rocks will fracture (if they have not done so already) and slip into a new position, releasing seismic energy as they do. Illustration by Don Vierstra.

As discussed below, the energy released during an earthquake, in the form of seismic waves, can be detected by an instrument called a seismograph. The record produced by a seismograph is seismogram. A worldwide network of seismic stations, or seismic observatories, provides records of seismic waves and their arrival times. Seismograms from at least three stations provide the data needed to locate the earthquake's origin.

Seismic Waves

Earthquakes release energy in the form of shock waves that vibrate the rocks composing Earth, both at the surface and within Earth's interior. These shock waves are called seismic waves, and they propagate in all directions away from the earthquake's origin, called the focus. The point on the earth's surface directly above the focus is known as the epicenter.

Earthquakes generate two general types of seismic waves: body waves and surface waves. Body waves travel through Earth—they penetrate its "body"—whereas surface waves travel along Earth's surface. Surface waves produce some of the ground movement we feel when we experience an earthquake, but they do not play a role in determining the location of an earthquake's epicenter.

Body waves, on the other hand, do provide information that can be used to determine an earthquake's focus and epicenter. There are two types of body waves: primary waves and secondary waves. They differ in the type of movement they cause within rock.

Primary waves, also called P waves, are compression waves. Compression waves have a push-pull motion (**Figure 4.2**) in the direction of the wave's travel. Sound waves are a good example of a compression wave.

Secondary waves, also called S waves, are shear waves. Shear waves have a sideways motion; they vibrate at right angles to the direction of propagation.

Seismographs also detect and record surface waves. There are two types of surface waves, based on the direction of ground movement: Love waves and Rayleigh waves. Love waves, also called LQ waves, are shear waves and move the ground from side to side. Rayleigh waves, also called LR waves, move the ground in a rolling motion. Both types of surface waves travel at similar velocities along the surface of the earth, and are the last seismic waves to arrive at a seismic station because they do not take shortcuts through the earth, as do body waves.

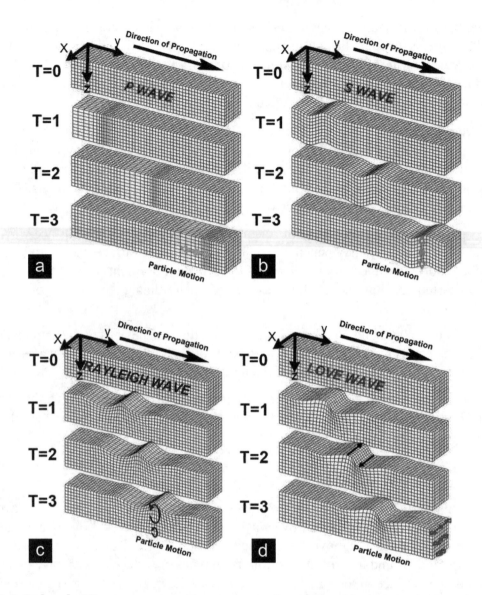

Figure 4.2 Seismic Waves. P waves (a), particles oscillate back and forth in the direction the wave is traveling. In S waves (b), particle motion is side to side. In Rayleigh waves (c), the waves travel along the surface in a vertical elliptical motion, and in Love waves (d), the waves travel along the surface in a side-to-side direction perpendicular to the direction of travel. Illustration by Larry Braile.

P waves and S waves provide seismologists with information about the location of earthquake's origin. An earthquake generates both P and S waves that travel in all directions away from the earthquake's focus; however, P and S waves travel at different velocities. Seismic velocity varies with the type of rocks the waves pass through and their state (temperature and pressure), but P waves travel faster than S waves. Therefore, the P wave is the first wave to arrive at a seismic station. The S wave follows within seconds or minutes of the arrival of the P wave, depending on distance from the epicenter.

Reading a Seismogram

A seismograph is the instrument used to detect the arrival of seismic waves at a seismic station. The recordings, called seismograms, show the passing seismic waves as squiggles against a time scale. The height of the squiggles from the baseline is known as amplitude. **Figure 4.3** is an example of a seismogram of an earthquake that occurred in Washington State as detected at an Oklahoma seismic station.

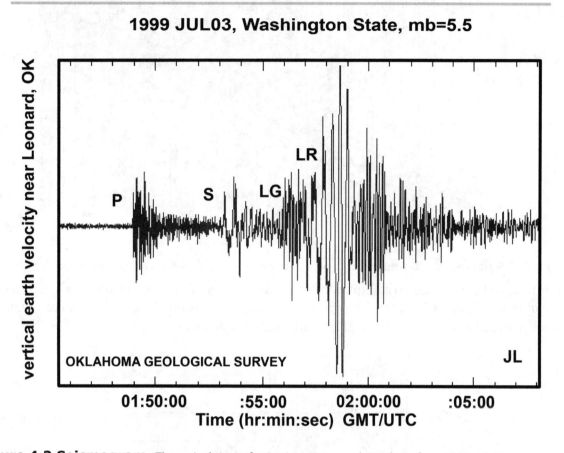

1999 JUL03, Washington State, mb=5.5

OKLAHOMA GEOLOGICAL SURVEY

JL

vertical earth velocity near Leonard, OK

01:50:00   :55:00   02:00:00   :05:00

Time (hr:min:sec) GMT/UTC

Figure 4.3 Seismogram. The arrival time of seismic waves are shown here for a 1999 seismic event. Note that LG waves are seismic waves that travel through continental crust. Illustration: Courtesy of Oklahoma Geological Survey.

The horizontal scale of the seismogram is time. All seismograms at seismic stations around the world are based on a standardized clock set to Greenwich Mean Time (GMT). At the far left of the seismogram are tiny squiggles with very low amplitudes that are the recordings of background vibrations present before the earthquake. These background vibrations include those caused by trucks, trains, heavy surf, construction equipment, and other localized sources. Most modern seismographs contain a damping mechanism and are placed in remote areas to keep background noise to a minimum.

The first earthquake vibrations to be detected are the P waves, followed by the S waves and then the surface waves. Their arrivals are noticeable by sudden changes in amplitude. Because P waves arrive first, most seismic stations will record P waves before S waves.

The time gap between the arrival of P and S waves increases with distance from an earthquake's focus. With every second that passes, P and S waves get farther apart and, therefore, so do their arrival times. **Figure 4.4** is a composite of seismograms recorded at many different seismic stations of a single earthquake that occurred near Wells, Nevada, in 2008. The growing gap between the arrival time of the P and S waves is clearly visible in this graph.

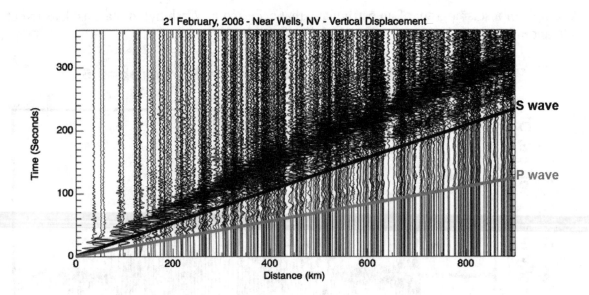

Figure 4.4 Seismogram Composite. Image: Courtesy of Charles Ammon, Pennsylvania State University.

The difference between the arrival times of P and S waves is often illustrated with a graph based on travel time (**Figure 4.5**). The graph gives seismologists a tool that can be used to calculate the distance between a seismic station and the earthquake's epicenter.

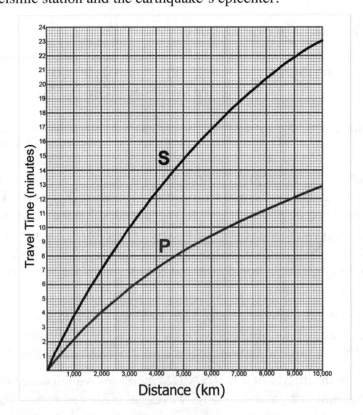

Figure 4.5 Travel Time Curves for P waves and S waves. This travel time graph allows seismologists to use the arrival times of P and S waves to calculate the distance from a seismographic station to an earthquake's epicenter. Illustration by Marie Hulett.

The gap in arrival times between P and S waves is used to calculate the distance between the seismic station and the epicenter and can be done using the travel time graph. Follow along on **Figure 4.6** below to see how this is done:

(1) *Subtract the arrival time of the P wave from the arrival time of the S wave to determine the time lag, known as S-P.* On **Figure 4.6**, the time lag has already been calculated to be 2.5 minutes.

(2) *Find an interval equivalent to the time lag on the y-axis of the travel time graph. Mark the start and end of that interval on a ruler or other straight edge.* On **Figure 4.6**, this would be the distance indicated by line *A*.

(3) *Find the gap between the P and S waves that equals the interval marked on the ruler or straight edge.* On **Figure 4.6**, this would be the gap indicated by the arrows.

(4) *Project down to the x-axis to read the distance that corresponds to this S-P interval.* On **Figure 4.6**, projecting down to the x-axis along line *c* gives us a distance of 1900 km.

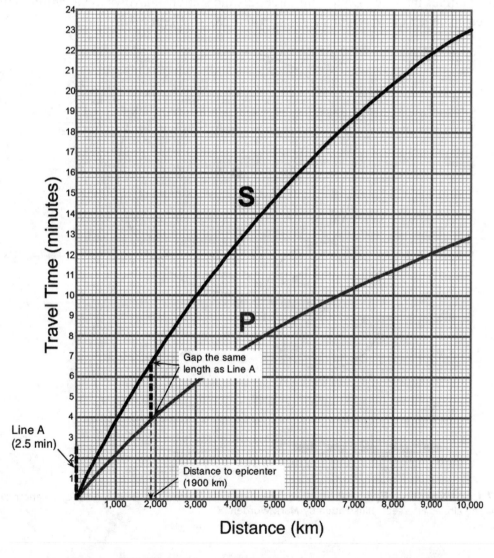

Figure 4.6 Distance to Epicenter. Illustration by Marie Hulett.

In reality, it is rare to find a seismologist using a travel time chart and straight edge to calculate distance to an earthquake's focus. Today, computers immediately make such calculations as soon as an earthquake is detected. Either way, however, this distance information is used to determine where the epicenter of an earthquake is located.

Locating the Epicenter

The location of an earthquake's focus is done through a technique called triangulation. This technique uses information from at least three seismic stations to determine where the earthquake occurred. **Figure 4.7** shows the locations of seismic stations A, B, and C. The S-P values for each seismic station have been calculated as described above. The epicenter distances have been plotted on a map by drawing a circle around each seismic station using the map scale provided to determine the radius for each circle.

Data from one seismic station provides a circle along which the epicenter is located; the epicenter could be anywhere on this circle as long as there is only data from one seismic station. Data from two seismic stations narrows down the location of the epicenter to only two possible points—where circles intersect. Data from three seismic stations narrows the location to the area where all three epicenter circles intersect, designated "X" in **Figure 4.7**.

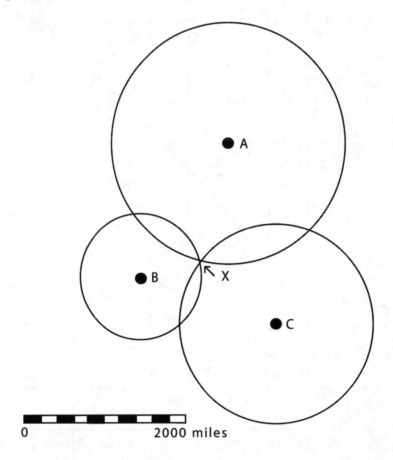

Figure 4.7 Locating the Epicenter. Three seismic circles narrow the location of the epicenter to the area where all the circles intersect. Illustration by Mark Worden.

Note that the circles do not always intersect perfectly; this happens when the earthquake's origin, the focus, is deep underground. The exact location of the focus is determined using the same data as is used to calculate an epicenter. However, determining the focus requires the use of spheres rather than circles, as illustrated in **Figure 4.8**. Once the distance between a seismic station and the earthquake's origin is calculated, that distance becomes the radius for a sphere, and the focus could be anywhere on the surface of the sphere that is beneath Earth's surface. When the data from a second seismic station is added, the focus can be anywhere at or beneath Earth's surface at which the surfaces of the two spheres intersect. Data from a third seismic station create a third sphere. There is only one possible point where all the surfaces of all three spheres would intersect, the earthquake's focus.

The epicenter is on the surface directly above the focus, but note that—if the focus is deep enough—the curvature of the spheres would mean that the triangulation circles from the same data (which are really the intersection of the spheres with the ground surface) might not actually meet on the surface. They will be close enough, however, to provide a good estimate of the location of the epicenter.

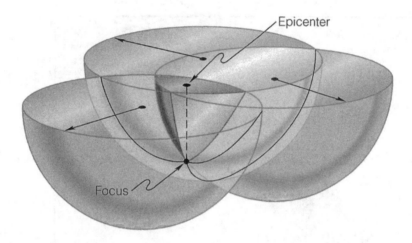

Figure 4.8 Locating the Focus. The focus of an earthquake can be pinpointed by drawing spheres from at least three seismic stations. The radius of each sphere is the distance from the seismic station to the focus, as calculated from the arrival times of P and S waves. The epicenter is the point on Earth's surface directly above the focus. Illustration by John J. Renton.

Wadati-Benioff Zone

Calculation of the epicenters and foci of earthquakes has given seismologists a detailed picture of where earthquakes happen and at what depth. The vast majority of earthquakes occur at plate boundaries, particularly at convergent and transform boundaries, providing evidence for plate tectonics theory.

Depth information has also supported plate tectonics theory. The evidence is based on both the depth and location of earthquakes near a subduction zone. An earthquake with a focus of less than 40 miles beneath Earth's surface is called a shallow-focus earthquake; an earthquake with a focus between 40 and 220 miles beneath Earth's surface is an intermediate-focus earthquake; earthquakes with a focus deeper than 220 miles beneath the surface are called deep-focus earthquakes. Most earthquakes are shallow-focus, but many intermediate- and deep-focus earthquakes occur near subduction zones.

During the 1940s, Hugo Benioff, a seismologist and professor at the California Institute of Technology, and Kiyoo Wadati, a seismologist at the Central Meteorological Observatory of Japan, were studying these intermediate- and deep-focus earthquakes. They discovered that earthquakes near subduction zones tend to occur along a plane (**Figure 4.9**). The plane, now called the Wadati-Benioff zone, represents the contact between the upper surface of the subducting oceanic plate and the rocks of the overlying asthenosphere.

The depth at which deep-focus earthquakes stop occurring is thought to be the depth at which the rocks of the subducting lithosphere become too ductile from the heat and pressure to produce earthquakes. Ductile rocks, such as those in the asthenosphere, respond to stress with plastic strain, permanently deforming rather than breaking. Since earthquakes are the result of brittle strain, no earthquakes can occur at such depths.

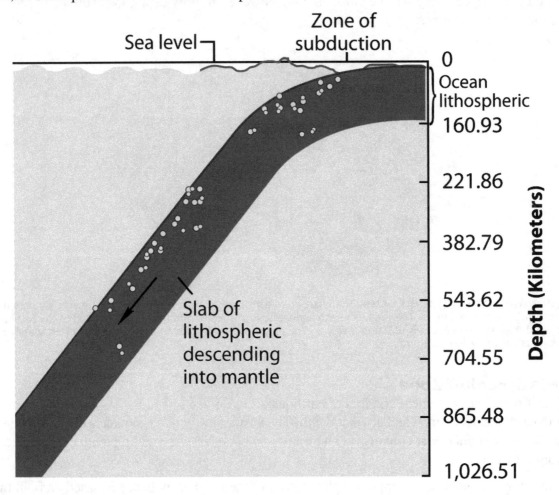

Figure 4.9 Wadati-Benioff Zone. Below the zones of subduction and dipping toward the continents at an angle of about 45 degrees is a plane called the Wadati-Benioff zone along which abundant earthquake foci are located. Modified from *Planet Earth* by John J. Renton. Copyright © 2002 by John J. Renton. Reprinted by permission of Kendall Hunt Publishing Company.

Lab Exercises

Lab Exercise #1: *Time Lag*

In this lab exercise, you will determine the distance to the epicenter of an earthquake. You will use the data to answer the questions at the end of this lesson.

Locating the epicenter of an earthquake to pinpoint accuracy is more easily done by computers than by hand. However, a close estimate of epicenter location can be made with a few simple calculations, beginning with time lag.

The first step in determining an epicenter is to calculate the time lag by subtracting the arrival time of the P waves from the arrival time of the S waves.

Instructions and Observations

Step 1: Use the seismogram in **Figure 4.3** to make these calculations. Each tick mark on the *x*-axis represents one minute.

1. What time in hours:minutes:seconds GMT did the P waves arrive?
   a. 01:48:00
   b. 01:49:00
   c. 01:50:00
   d. 01:51:00

2. What time in hours:minutes:seconds GMT did the S waves arrive?
   a. 01:51:30
   b. 01:52:00
   c. 01:53:10
   d. 01:54:20

3. What is the time lag (S-P) in hours:minutes:seconds?
   a. 00:02:00
   b. 00:03:20
   c. 00:03:30
   d. 00:04:10

Step 2: Now use **Figure 4.10** below (which is a portion of the travel time chart **Figure 4.5**) to determine the distance between the seismic station and the epicenter in kilometers.

4. What is the distance in kilometers between the seismic station and the epicenter?
   a. 2400 km
   b. 2600 km
   c. 2800 km
   d. 3000 km

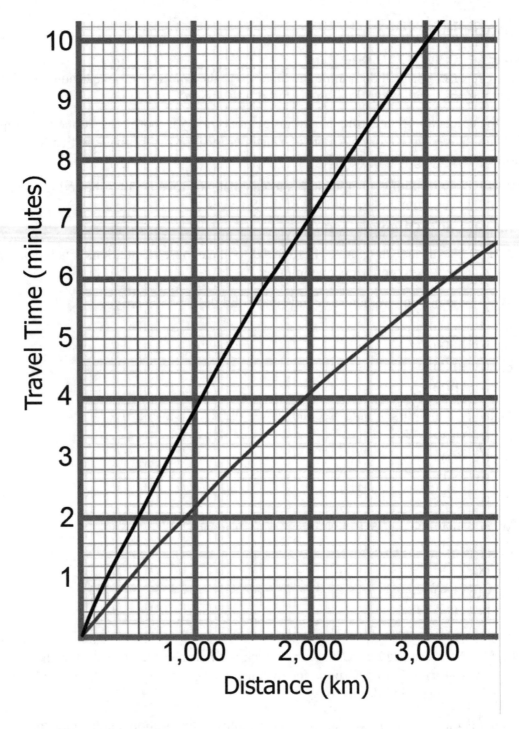

Figure 4.10 Portion of Travel Time Chart. Illustration by Marie Hulett.

Lab Exercise #2: *Locating the Epicenter of an Earthquake*

In this lab exercise, you will locate the epicenter of an earthquake. A minimum of three seismic stations is necessary to locate an epicenter. As discussed earlier in this lesson, this method of determining the epicenter of an earthquake is known as triangulation.

Below is an illustration (**Figure 4.11**) that shows seismogram records for a hypothetical earthquake as recorded in three cities. P waves are the first wave to appear; S waves appear next. The waves appear on the graph when there is a sudden change in the height of the squiggles.

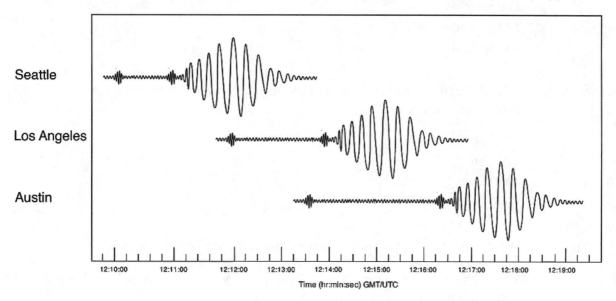

Figure 4.11 Seismogram Records for Three Cities. Illustration by Susan Wilcox.

Follow the instructions below to locate the epicenter of an earthquake based on the seismograms in **Figure 4.11**. You'll begin by determining the time lag between the arrival of the P and S waves at each seismic station, then use the travel time chart in **Figure 4.10** to determine the distance between each seismic station and the epicenter.

Once you've answered the following time lag and distance questions, use the triangulation method described below to locate the epicenter. Feel free to photocopy **Figure 4.12** or to download a copy of this map from the online supplement so that you can make multiple attempts, if necessary. Submit your work as directed by your instructor.

Instructions and Observations

Step 1: What is the time lag for noted at the Seattle, Washington, seismic station?

    a. 45 seconds
    b. 50 seconds
    c. 55 seconds
    d. 65 seconds

Step 2: What is the distance between the Seattle seismic station and the earthquake's epicenter?

    a. 600 km
    b. 700 km
    c. 800 km
    d. 900 km

Step 3: What is the time lag for noted at the Los Angeles, California, seismic station?

    a. 100 seconds
    b. 120 seconds
    c. 140 seconds
    d. 160 seconds

Step 4: What is the distance between the Los Angeles seismic station and the earthquake's epicenter?

    a. 1000 km
    b. 1100 km
    c. 1300 km
    d. 1500 km

Step 5: What is the time lag for noted at the Austin, Texas, seismic station?

    a. 120 seconds
    b. 140 seconds
    c. 165 seconds
    d. 180 seconds

Step 6: What is the distance between the Austin seismic station and the earthquake's epicenter?

    a. 1950 km
    b. 2200 km
    c. 2550 km
    d. 2850 km

Use the information obtained above and the map below (**Figure 4.12**) to locate the epicenter.

Step 7: Using the scale on the map in **Figure 4.12**, put one point of your drawing compass on 0 km and the other compass point on the distance calculated in step 2 above (for Seattle).

Step 8: Carefully lift the compass from the scale (so as not to change the distance between the compass points) and place one compass point on Seattle. Place the other compass point down and draw an arc across the map (there will not be room for a complete circle as shown in **Figure 4.7**).

Step 9: Repeat steps 7 and 8 above to draw an arc the proper distance from Los Angeles.

Step 10: Repeat steps 7 and 8 above to draw an arc the proper distance from Austin.

Step 11: Draw a small triangle at the point where the three arcs intersect. This is the epicenter of the earthquake.

                                           Lesson 4/Earthquakes and Seismology

Step 12: What city is nearest to the epicenter?

a. Helena
b. Bismark
c. Boise
d. Cheyenne

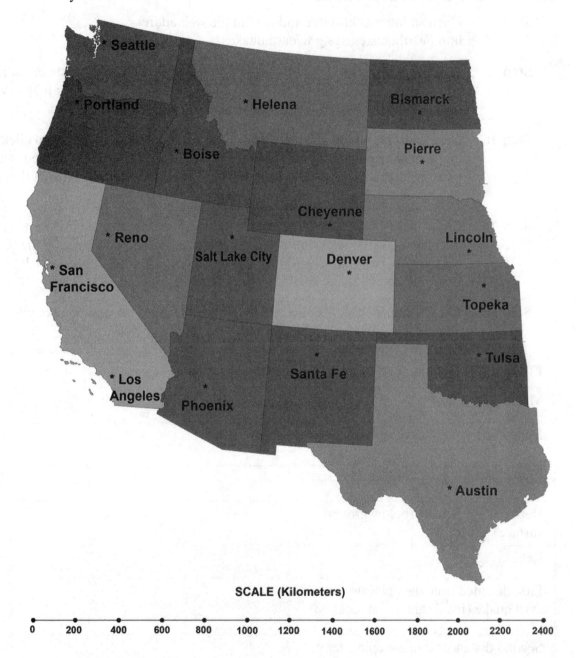

Figure 4.12 Western Part of the United States. Illustration by Marie Hulett.

Lab Exercise #3: *USGS Earthquake Hazards Program*

In this lab exercise, you will go to the United States Geological Survey (USGS) website "Earthquake Hazard Program" where you will investigate how to look up current earthquake activity occurring in the continental United States.

Instructions and Observations

Step 1: Open an Internet browser and type in the web address http://earthquake.usgs.gov/earthquakes

Step 2: Find the map of the United States on the web page, locate the state of California, and double click on California. The USGS California-Nevada Fault Map will open.

Step 3: Drag your mouse (displayed as a little hand icon) across the map and click on the largest square on the map. If there are several "largest squares" of the same size, pick one that appeals to you. Note the approximate magnitude size of the square you have chosen as indicated on the map legend.

Step 4: Click on the square you have chosen until you see the Earthquake Details. Check the magnitude to make sure it is at least the magnitude you noted in Step 3; this ensures you have information for the square you have chosen, especially when it is surrounded by many smaller earthquakes.

Step 5: The Earthquake Details data shows the most recent information about that particular event. Use the table below to record your findings.

Earthquake Details	Results
Magnitude	
Date-Time	
Location (GPS coordinates)	
Depth of earthquake (below ground surface)	
Region	
First distance from the epicenter of the earthquake (list distance and location)	
Second distance from the epicenter of the earthquake (list distance and location)	
Event ID	

Online Activities

Per your instructor's directions, go to the online supplement for this lab and complete the activities assigned. Viewing the online videos will help you to complete the quiz.

Quiz

Multiple Choice

Questions 1 through 4 are based on **Lab Exercise #1:** *Time Lag.*

1. Record your answer from **Lab Exercise #1, Step 1, Question 1**. What time in hours:minutes:seconds GMT did the P waves arrive?
   a. 01:48:00
   b. 01:49:00
   c. 01:50:00
   d. 01:51:00

2. Record your answer from **Lab Exercise #1, Step 1, Question 2**. What time in hours:minutes:seconds GMT did the S waves arrive?
   a. 01:51:30
   b. 01:52:00
   c. 01:53:10
   d. 01:54:20

3. Record your answer from **Lab Exercise #1, Step 1, Question 3**. What is the time lag (S-P) in hours:minutes:seconds?
   a. 00:02:00
   b. 00:03:20
   c. 00:03:30
   d. 00:04:10

4. Record your answer from **Lab Exercise #1, Step 2, Question 4**. What is the distance in kilometers between the seismic station and the epicenter?
   a. 2400 km
   b. 2600 km
   c. 2800 km
   d. 3000 km

Questions 5 through 11 are based on **Lab Exercise #2:** *Locating the Epicenter of an Earthquake.*

5. Record your answer from **Lab Exercise #2, Step 1**. What is the time lag for noted at the Seattle, Washington, seismic station?

    a. 45 seconds
    b. 50 seconds
    c. 55 seconds
    d. 65 seconds

6. Record your answer from **Lab Exercise #2, Step 2**. What is the distance between the Seattle seismic station and the earthquake's epicenter?

    a. 600 km
    b. 700 km
    c. 800 km
    d. 900 km

7. Record your answer from **Lab Exercise #2, Step 3.** What is the time lag for noted at the Los Angeles, California, seismic station?

    a. 100 seconds
    b. 120 seconds
    c. 140 seconds
    d. 160 seconds

8. Record your answer from **Lab Exercise #2, Step 4**. What is the distance between the Los Angeles seismic station and the earthquake's epicenter?

    a. 1000 km
    b. 1100 km
    c. 1300 km
    d. 1500 km

9. Record your answer from **Lab Exercise #2, Step 5**. What is the time lag for noted at the Austin, Texas, seismic station?

    a. 120 seconds
    b. 140 seconds
    c. 165 seconds
    d. 180 seconds

10. Record your answer from **Lab Exercise #2, Step 6**. What is the distance between the Austin seismic station and the earthquake's epicenter?

    a. 1950 km
    b. 2200 km
    c. 2550 km
    d. 2850 km

11. Record your answer from **Lab Exercise #2, Step 12**. What city is nearest to the epicenter?

    a. Helena
    b. Bismark
    c. Boise
    d. Cheyenne

12. The focus of an earthquake is a point

    a. on the earth's surface where the initial release of seismic energy occurs.
    b. on the earth's surface where seismic waves are reflected downward.
    c. within the earth where seismic waves are reflected downward.
    d. within the earth where the initial release of seismic energy occurs.

13. The epicenter of an earthquake is a point

    a. on the earth's surface directly above the initial release of seismic energy.
    b. within the earth directly beneath the initial release of seismic energy.
    c. on the earth's surface where the initial release of seismic energy occurs.
    d. within the earth where the initial release of seismic energy occurs.

14. Which of the following seismic waves has the same wave form as a sound wave?

    a. surface wave
    b. P wave
    c. S wave
    d. Love wave

15. Which of the following seismic waves is the first to arrive at a seismic station?

    a. surface wave
    b. P wave
    c. S wave
    d. Love wave

16. Which of the following seismic waves is the last to arrive at a seismic station?

    a. Love wave
    b. P wave
    c. S wave
    d. body wave

17. Which seismic wave vibrates side-to-side relative to its direction of travel?

    a. surface wave
    b. P wave
    c. S wave
    d. Rayleigh wave

18. The two kinds of body waves are
    a. surface waves and P waves.
    b. Love waves and P waves.
    c. P waves and S waves.
    d. Love waves and S waves.

19. The following seismic waves are used to locate earthquake epicenters:
    a. surface waves and P waves.
    b. Rayleigh waves and P waves.
    c. P waves and S waves.
    d. surface waves and S waves.

20. Which of the following statements is correct?
    a. P-wave amplitude is larger the S-wave amplitude.
    b. S-wave amplitude is larger than P-wave amplitude.
    c. P-wave amplitude and S-wave amplitude are about the same.
    d. P-wave amplitude is larger than all the other seismic waves recorded on a seismogram.

21. Determining the distance along the earth's surface from a seismic station to the epicenter of an earthquake is based on the difference in arrival times between
    a. the first P wave and the first S wave.
    b. the first P wave and the first Love wave.
    c. the first S wave and the first Love wave.
    d. all three seismic waves.

22. To locate the epicenter of an earthquake, it takes
    a. one seismic station.
    b. at least two seismic stations.
    c. at least three seismic stations.
    d. at least four seismic stations.

Short Answer
Questions 1 through 8 are based on **Lab Exercise #3:** *USGS Earthquake Hazards Program.*

1. Record your answer from **Lab Exercise #3, Step 5**, Magnitude.

2. Record your answer from **Lab Exercise #3, Step 5**, Date-Time.

3. Record your answer from **Lab Exercise #3, Step 5**, Location (GPS coordinates).

4. Record your answer from **Lab Exercise #3, Step 5**, Depth of earthquake (below ground surface).

5. Record your answer from **Lab Exercise #3, Step 5**, Region.

6. Record your answer from **Lab Exercise #3, Step 5**, First distance from the epicenter of the earthquake (list distance and location).

7. Record your answer from **Lab Exercise #3, Step 5**, Second distance from the epicenter of the earthquake (list distance and location).

8. Record your answer from **Lab Exercise #3, Step 5**, Event ID.

Lesson 4/Earthquakes and Seismology

MINERALS

Lesson 5

AT A GLANCE

Purpose

Learning Objectives

Materials Needed

Overview

Lab Exercise

Online Activities

Quiz

Purpose

This laboratory lesson will familiarize you with the physical properties of minerals, how the physical properties are used to identify minerals, and the economic importance and practical uses of minerals.

Learning Objectives

After completing this laboratory lesson, you will be able to:
- Describe the economic importance of minerals and some of their practical uses.
- Describe the physical properties that identify and distinguish minerals.
- Identify minerals by means of distinguishing among their unique set of physical properties (hardness, fracture surfaces, streak, color, and so on).

Materials Needed

The Mineral Identification kit, which is part of your lab materials, is required for this exercise.

- ❑ Common household lemon juice, lime juice, or vinegar
- ❑ 10X-magnifying hand lens (in lab kit)
- ❑ Porcelain streak plate (in lab kit)
- ❑ Glass plate (in lab kit)
- ❑ Wire nail (in lab kit)
- ❑ Magnet (in lab kit)
- ❑ Ten mineral specimens (in lab kit)
- ❑ White sheet of paper
- ❑ Old copper penny
- ❑ White sheet of paper

Overview

Contrary to popular belief, rocks and minerals are not the same thing. A mineral is defined as any substance that meets these requirements:

- It is naturally occurring. Man-made substances do not qualify as minerals.
- It is inorganic. Substances that are or have been alive do not qualify as minerals.
- It has a definite chemical composition. It always has the same composition of elements.
- It has an orderly and predictable atomic structure. The atoms composing a mineral arrange themselves in a characteristic form, called the mineral's crystalline structure.
- It has physical properties that are characteristic of that substance.

If a substance meets all five of these requirements, it is a mineral. If it doesn't, it is not a mineral no matter how much it may look like one. This is a different definition of mineral than the one used by nutritionists, who have labeled as "minerals" certain elements, like calcium, magnesium, and iron, that are necessary for life processes to take place. While these substances are ultimately derived from geologic minerals, they are not minerals in the geologic sense.

Rocks have a less complicated definition. Most rocks are composed of minerals, although some contain other substances as well; one may think of most rocks as solid mixtures of minerals. Lessons 6, 7, and 8 will focus on the various types of rocks formed by Earth's minerals. This exercise will feature some of the common rock-forming minerals and a few others of economic importance. What you will learn about minerals in this lesson will contribute greatly toward your successful completion of the later lessons on rock identification.

The Economic Importance and Practical Uses of Minerals

The average American uses more than 40,000 pounds of newly mined and extracted minerals every year, just in the process of living our lives. We are unaware of many of these minerals. Consider that about 15 different minerals contribute to the making of a car, 35 different minerals contribute to a television, 30 minerals contribute to a computer, and over 40 different minerals contribute to a telephone, and it becomes abundantly clear the role minerals play in our lifestyles and standard of living. A look around the room will reveal few objects that don't have some mineral as their original source. The glass in the windows, the drywall beneath the paint or wallpaper, the copper wiring carrying electricity, the metal in the chair, and even the chips running the computer all contain minerals that have been modified to serve a useful purpose.

Humans have been mining minerals for thousands of years. Consumption of many of these minerals has increased exponentially over the past several decades. Our modern society depends upon the major minerals that are extracted from Earth's crust through mining, such as gold, silver, platinum, and diamond. Other types of rocks and minerals provide the infrastructure of our current civilization. For example, sheetrock and plaster are produced from mineral called gypsum and are used as wallboard for the interior walls in homes. Silica sand, which is about 95 percent quartz, is used for making glass, ceramics, and microchips. The table salt that seasons our food and rock salt that melts ice from our driveways are actually the mineral halite, also known as sodium chloride ($NaCl$). The halite comes from ancient seas or lakebeds that dried up, leaving salt deposits that are mined for use by human beings.

Figure 5.1 lists some other commonly found minerals and their economic uses.

Mineral	Economic Uses
Calcite	As primary mineral in limestone, used as fertilizer, soil amendment, as the Portland cement in concrete; main component of marble and travertine used as building stone; main component in Mexican onyx used by sculptors
Chalcopyrite	Source of copper
Feldspar	Major ingredient in manufacture of porcelain,glass, glazes; semiprecious gems, e.g., moonstone, sunstone, amazon stone
Galena	Chief ore of lead
Garnet	Abrasive powder; emery cloth; sandpaper; gemstone
Graphite	Lubricant for machinery; pencil lead; used in electrical industry and in the production of batteries
Gypsum	Sheetrock; plaster; fertilizers; filler in paper
Halite	Table salt; source of sodium and chorine; used for deicing roadways; old salt mines used as storage sites
Hematite	Major source of iron; minor use as pigment
Limonite	Minor ore of iron; pigment
Magnetite	Ore of iron
Mica	Used as electrical insulation; isinglass, the "windows" in wood-burning stoves
Olivine	Used as a refractory (substance resistant to heat); source of magnesium; variety Peridot is semiprecious gem
Pyrite	Minor source of sulfur to make sulfuric acid
Quartz	Source of silica for glass, grit on sand paper; source of silicon for computer chips; varieties are semiprecious gems amethyst, tiger's eye, and onyx

Figure 5.1 Economic Uses of Minerals. Illustration by Susan Wilcox.

The importance of minerals runs even deeper, however. As noted above, minerals supply elements necessary for the chemical reactions of life to occur, without which the paper in this book, the wood frames of most homes, and the person reading this lab manual could not exist. Even the atoms in the air were once trapped in the minerals of a young Earth. It is no exaggeration to state that virtually everything one sees, hears, and touches is there, one way or another, courtesy of minerals.

Mineral Groups

Geologists have identified more than 3,800 mineral species, of which only fifteen or twenty are considered common. The chemical composition and crystalline structure of a mineral determine its physical properties—its color, its hardness, its shape, its feel, and how it reflects or refracts light.

Although composed of even smaller units called atoms, minerals are the smallest units that can be seen with the unaided eye or a low-magnification hand lens. The consistency in a mineral species' atomic arrangement from specimen to specimen, regardless of where the mineral crystallized, is one of nature's amazing phenomena.

Silicate Minerals

Minerals can be categorized into different groups based on their composition and structure. The most abundant group, by far, is the silicates. Rocks composed of silicate minerals comprise about 90 percent of Earth's crust and nearly 100 percent of Earth's mantle.

The main building block of a silicate mineral is the silicon-oxygen tetrahedron in which four oxygen atoms surround a silicon atom (SO_4) in a tetrahedral structure (**Figure 5.2**).

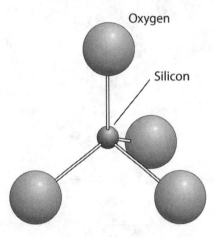

Figure 5.2 The Oxygen-Silicon Tetrahedron. Illustration by John J. Renton.

Silicon-oxygen tetrahedra can stand alone or link together in a variety of ways that affect the physical properties of the minerals they form. Each type of arrangement, called a silicate structure, forms a distinctive subgroup of the silicate group of minerals (**Figure 5.3**). These structures determine some of the properties of the minerals in each group.

- The olivine group of minerals contains silicon-oxygen tetrahedra that do not link together; this is called an isolated tetrahedra structure.

- The pyroxene group of minerals contains silicon-oxygen tetrahedra that link together to form single chains, recognizable by the way these minerals cleave. Cleavage is discussed later in this lesson.

- The amphibole group of minerals contains tetrahedra that link together to form double chains, producing a different type of cleavage than the pyroxene group.

- The mica group of minerals contains tetrahedra that link together to form sheets. As discussed later in this lesson and in subsequent lessons, this sheet structure accounts for some of the most recognizable properties of rocks formed by mica minerals.

- Quartz and members of the feldspar group of minerals have a framework structure in which the silicon-oxygen tetrahedra bond in all directions. Feldspar has cleavage whereas quartz does not.

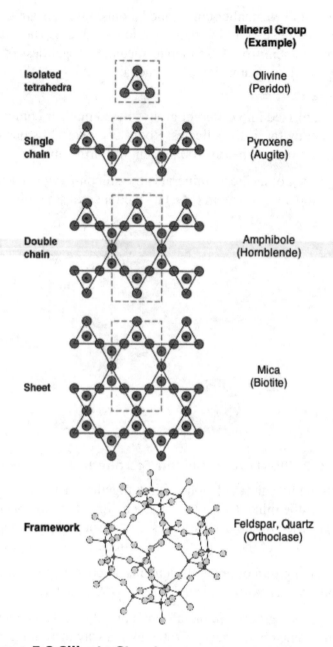

Mineral Group (Example)

Isolated tetrahedra — Olivine (Peridot)

Single chain — Pyroxene (Augite)

Double chain — Amphibole (Hornblende)

Sheet — Mica (Biotite)

Framework — Feldspar, Quartz (Orthoclase)

Figure 5.3 Silicate Structures. Illustration by John J. Renton.

In addition to the different types of silicate structures, silicate minerals also vary by composition. While they all contain silicon-oxygen tetrahedra, they contain other atoms as well that help to hold the structure together.

- Minerals in the olivine, pyroxene and amphibole group contain the additional atoms, iron or magnesium, so they are called ferromagnesian minerals. They are also called mafic minerals. The iron and magnesium make mafic minerals dark in color.

- Minerals in the feldspar group contain the additional atoms, calcium, sodium, or potassium. Feldspars, as well as quartz, are called felsic minerals. The calcium, sodium or potassium atoms give felsic samples with few impurities a light color, but impurities can quickly darken a sample so color is always not a reliable indicator of whether a sample is mafic or felsic.

- Quartz contains no additional atoms in its structure; it is composed entirely of silicon and oxygen.

- Mica can be either mafic or felsic. Biotite is mica with a mafic composition; muscovite is mica with a felsic composition.

Nonsilicate Minerals

Besides the silicon-oxygen tetrahedra, the atoms of some other elements can form the base for a group of minerals either as single atoms or as part of a group of atoms.

- Carbonates are minerals that contain the carbonate group, a carbon atom with three oxygen atoms (CO_3). Carbonates are soft minerals, as will be described below, and are distinctive in that they react with acid.

- Oxides are minerals in which an element has combined with oxygen. Magnetite and hematite are both iron oxides (Fe_2O_3); an example of an iron oxide is rust.

- Sulfates are minerals that contain the sulfate group, four oxygens and a sulfur atom (SO_4).

- Sulfides are minerals formed when a single sulfur atoms combines with another element. When iron combines with sulfur, the result is pyrite. Lead and sulfur produce galena. Zinc and sulfur produce sphalerite.

- Halides are minerals formed when an element combines with chlorine or fluorine; for instance, halite (common table salt) is sodium and chlorine (NaCl) and fluorite is calcium and fluorine (CaF_2).

- Native elements are those that do not combine with other atoms. Examples of native elements are gold, silver, platinum, and copper.

There are other groups as well, but Earth's most common minerals are included in the group listed above.

Physical Properties of Minerals

The different structures and compositions of Earth's minerals produce different properties. Because the atomic structure of a mineral species is always the same, most of its physical properties are relatively constant and may be used for the mineral's identification. The more diagnostic physical properties (those used to make an identification) of minerals are color, luster, streak, hardness, crystal form, cleavage/fracture, specific gravity, taste, smell, reaction to acid and magnetism.

Remove Sample A from the bag marked Lab #5 Mineral Samples from your lab kit. You are going to test this sample to determine each of the physical properties described below, and then use what you discover to make your first mineral identification.

Color

Color is the most obvious physical characteristic of any mineral specimen, but it is the least reliable property when it comes to mineral identification. For some minerals, it provides a useful clue for identification. For other minerals, color can be misleading due to impurities. For example, quartz may be colorless, or may be one of a variety of colors such as white, rose, gray, yellow, or brown. Mineral surfaces exposed to the elements may show a different color and/or transparency than freshly exposed surfaces, so be aware of the factors affecting color.

However, for some mineral groups, a common color range can be used as a clue to aid in mineral identification. In particular, mafic minerals are dark green or black in color while felsic minerals tend to be light. For example, plagioclase feldspar, the type of feldspar containing calcium, is often a whitish-gray color, and orthoclase feldspar, often called potassium feldspar, often has a pink color. Biotite, the mafic variety of mica, is always black or dark brown. Muscovite, the felsic variety of mica, ranges from colorless to white and light browns. Hornblende, a member of the mafic amphibole group, is typically dark green-to-black.

Given the above, when identifying minerals, it is best to note the color as a possible clue, but rely more on other physical properties to complete your identification.

Take a look at Sample A and note its color. _____

Luster

Luster is the appearance of light reflected from a fresh mineral surface, one that hasn't been worn down or weathered. The first question to ask is whether a sample has a metallic or nonmetallic luster (see **Figure 5.4**). A mineral with a metallic luster reflects light like a metal such as gold (polished or brushed gold jewelry), brass, copper, silver, or stainless steel (a knife blade or stainless kitchen appliance). A poor metallic luster, opaque but reflecting little light, is called a submetallic luster.

Note that metallic does not imply "shininess" or anything about color. For example, the mineral biotite is black and shiny, but it has a nonmetallic luster. The distinction, however, between metallic and nonmetallic luster is not always as clear-cut as it might appear to be. For some mineral specimens, it helps to hold them up to a light source; minerals with a metallic luster are opaque, so you cannot see light passing through even thin edges, whereas light readily shows through the thin edges or through the thin sheets of a biotite sample, which is nonmetallic.

Luster varies over a wide range, and there is no rigid boundary for using the various terms to describe luster. Luster can even vary widely within a particular mineral species. For example, gypsum can have a vitreous to pearly, silky, waxy, or sometimes earthy luster. Descriptions of nonmetallic luster include:

> resinous (looks like a dried resin, plastic or glue)
>
> pearly (reflects light like a pearl)
>
> earthy or dull (nonreflective, like dried mud)
>
> silky (fibrous structure like silk cloth)
>
> waxy or greasy (like a wax candle or grease)
>
> glassy (also called vitreous)

Garnet crystals and sphalerite have a resinous luster. Muscovite mica has a pearly luster. Kaolinite and other clay minerals have a dull or earthy luster, asbestos has a silky luster, chalcedony has a waxy luster, cordierite has a greasy luster, and quartz has a glassy luster.

Metallic (gold)

Vitreous (rose quartz)

Resinous (garnet crystals)

Waxy (chalcedony)

Pearly (muscovite mica)

Dull-earthy (barite)

Figure 5.4 Lusters. The minerals above display specific metallic or nonmetallic lusters. Photos: Metallic, Shutterstock 51465025, credit Matthew Benoit; Vitreous, Shutterstock 58655551, credit optimarc; Resinous, Shutterstock 27787174, credit Nikolai Pozdeev; Waxy, Shutterstock 24152110, credit dmitriyd; Pearly, Shutterstock 56061226, credit Tyler Boyes; Dull-earthy, Shutterstock 2188460, credit Manamana.

Compare Sample A to the luster examples in **Figure 5.4**. What type of luster does Sample A appear to have? _____

Streak

Streak is the color of a mineral's powder. Streak is determined by drawing the mineral specimen across a piece of unglazed porcelain, called a streak plate, and rubbing the streak with your finger. Some minerals have a streak that is the same color as the color of the sample. Others have a streak that differs markedly in color from the hand specimen. While, as noted above, the color of a sample may vary, the color of a streak is consistent from sample to sample. For instance, magnetite always has a black streak, and hematite always has a reddish brown streak, regardless of the sample's surface appearance. This makes streak a particularly useful property in mineral identification.

The streak of minerals with a metallic luster is especially diagnostic. A streak test would be very helpful in distinguishing between real gold, which has a yellow streak, and fool's gold (pyrite), which has a greenish-black streak. Unfortunately, because many nonmetallic minerals have white or pastel streaks, a streak test may not be useful unless you have access to a brown or black streak plate. Also, minerals that are harder than a streak plate will not leave a powered streak but instead scratch the streak plate into a powder.

Using the streak plate from your lab kit, conduct a streak test on Sample A by drawing it across the plate. Can you see a streak? _____

Hardness

Hardness describes a mineral's ability to resist scratching or abrasion, and is dependent upon the type of atomic bonds. Hardness is measured on a scale of one to ten, called Mohs scale of hardness, which lists ten common minerals ranked in order of increasing relative hardness from talc (1) to diamond (10) (**Figure 5.5**). Every mineral has a hardness that falls somewhere on this scale.

Mohs Scale of Hardness		Hardness of Some Common Objects	
Hardness	Mineral	Hardness	Object
10	Diamond		
9	Corundum		
8	Topaz		
7	Quartz		
6	Feldspar	5.5	Glass plate, steel knife
5	Apatite	4.5	Wire, nail
4	Fluorite	3.5	Copper penny
3	Calcite	2.5	Fingernail
2	Gypsum		
1	Talc		

Figure 5.5 Mohs Scale of Hardness. Illustration by Susan Wilcox.

Even though each mineral on the hardness scale has an assigned number, the scale is relative, not absolute. Fluorite is a 4 on the scale, but it is not twice as hard as gypsum (number 2), nor is it half as hard as topaz (number 8). Mohs hardness scale simply compares the hardness of one mineral to that of another.

The hardness of a mineral sample is determined through trial and error using one of two methods:

- Using a substance with a known hardness like a wire nail (4.5) or a mineral on the Mohs scale like quartz (7) to scratch the sample.

- Using the mineral sample itself to try to scratch a substance of known hardness.

Higher-numbered minerals will scratch lower-numbered minerals, but lower-numbered minerals cannot scratch higher-numbered minerals, so you can use a mineral sample with a known identity to do a scratch test on an unknown mineral sample. In addition, a mineral sample will scratch an object with a lower hardness value but not one with a higher hardness value. The hardness of some objects commonly used in mineral hardness tests is also listed in **Figure 5.5**.

Unless a mineral sample feels uncommonly hard or soft to you, a good starting place is in the middle of the scale with a scratch test on the glass plate (one is included in your lab kit). **In making hardness tests on a glass plate, however, do not hold the glass plate in your hands; keep the glass plate firmly on the table top before applying pressure with the mineral specimen.** If you think that you have made a scratch on the glass, try to rub the scratch off. What appears to be a scratch may be only some of the mineral that has rubbed off on the glass. All minerals of hardness 5.5 and higher will scratch the glass plate.

Once you've determined if the hardness of your mineral sample is above or below 5.5, you can do other scratch tests. A steel knife blade will scratch all minerals with hardness of 5.5 or less, whereas a fingernail will scratch all minerals with hardness of 2.5 or less. For example, you can scratch calcite (hardness 3) with a copper penny (hardness 3.5) but you cannot scratch the calcite with your fingernail (hardness 2.5). You would record this hardness during your mineral identification as 2.5–3.5. Any mineral whose hardness falls within this range has the possibility of being your sample; checking other properties will help you to narrow down the possibilities.

Do a scratch test on Sample A to determine its hardness. What hardness range does it have on Mohs scale?

   a. 1–2.5
   b. 2.5–3.5
   c. 3.5–5.5
   d. > 5.5

Crystal Form

If a mineral forms in an environment where its growth is unimpeded, it may develop smooth symmetrical faces that, together, outline a specific geometric shape. These symmetrical geometric forms are called crystals. The specific shape is known as a crystal form. A mineral has one, or sometimes more, characteristic crystal forms. Its crystals will only take those forms.

A well-formed mineral crystal is one of the most beautiful objects produced in nature. Perfect crystals in nature, however, are the exception rather than the rule and usually form only under special conditions where there is open space for them to grow while the mineral is forming.

Many mineral samples are broken pieces of larger crystals, so mineral samples don't always show crystal form. But if crystals are visible, they may be used for mineral identification process

Halite, also known as table salt, is an example of a mineral with a cubic crystal form; salt crystals are always cubes. A dodecahedron is one of the crystal forms of diamonds. The mineral magnetite crystallizes as an octahedron, and the mineral quartz, crystallizes as a hexagonal prism that ends in a six-sided pyramid (**Figure 5.6**).

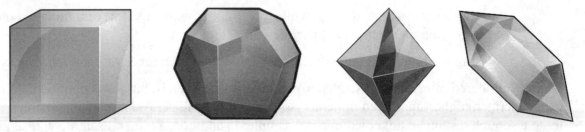

Figure 5.6 Common Crystal Forms. Crystal forms from left to right: cubic, dodecahedral, octahedral, hexagonal. From *Planet Earth* by John J. Renton. Copyright © 2002 by John J. Renton. Reprinted by permission of Kendall Hunt Publishing Company.

Crystal form is a diagnostic property for a mineral sample only if the crystals are visible. Some hand samples are single large crystals. Others have smaller crystals visible to the naked eye on the surface of the sample, and many more samples have crystals visible only under a magnifying glass or microscope.

Examine Sample A with and without the magnifying glass contained in your lab kit. Can you see any crystals? _____

If so, does it have one of the crystal forms described above? _____

Cleavage

Cleavage is another diagnostic property of minerals. Cleavage is the tendency of some minerals to split along parallel planes of weakness within their structure. Each plane is a cleavage direction or cleavage plane. Not all minerals exhibit cleavage and if cleavage is exhibited, the quality of cleavage in a mineral can range from excellent to poor.

To understand cleavage planes better, set a small box or a textbook on a table and note the parallel directions:

- Top and bottom are parallel to each other, and so they are one parallel direction, or one plane.

- The two long sides are parallel to each other, and so they form another parallel direction or plane.

- The two short sides are parallel to each other, and form a third parallel direction or plane.

Thus, even though the box or book has six sides, it has three planes. In minerals, however, the planes are rarely as obvious as they are with the book or the box.

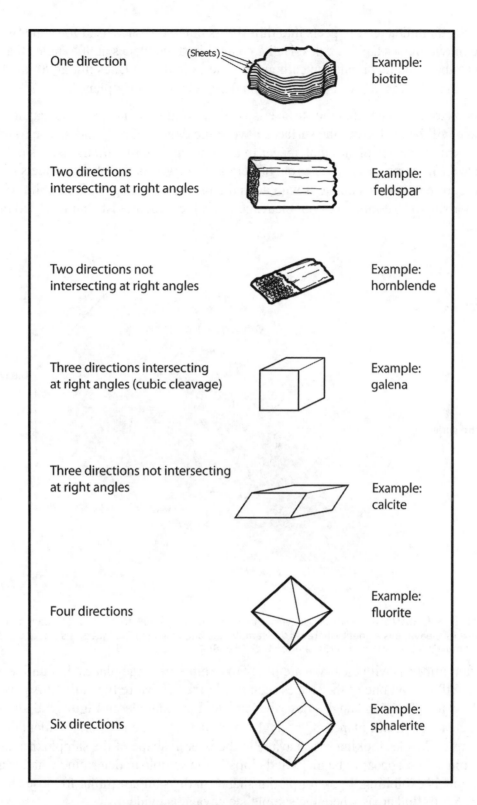

One direction   (Sheets)   Example:
                                                                            biotite

Two directions   Example:
intersecting at right angles   feldspar

Two directions not   Example:
intersecting at right angles   hornblende

Three directions intersecting   Example:
at right angles (cubic cleavage)   galena

Three directions not intersecting
at right angles   Example:
                                                                           calcite

Four directions   Example:
                                                                           fluorite

Six directions   Example:
                                                                           sphalerite

Figure 5.7 Mineral Cleavage. Illustration by Don Vierstra.

Figure 5.7 shows the cleavage planes characteristic of the silicate groups. Pyroxenes, the single-chain silicates, cleave in two directions at 90° angles to each other. Amphiboles, the double-chain silicates, cleave in two directions but **NOT** at 90° angles; the cleavage angle for opposite corners is 60° and 120°. The mica group with its sheet structure cleaves in only one plane; the cleavage is so perfect that mica readily peels into thin sheets that were once used for windows. Feldspars cleave in two planes at 90° angles. Halite cleaves in three planes at 90° angles that can form a perfect cube. Calcite, or calcium carbonate, cleaves in three planes **not** at 90° angles. Fluorite is a mineral that cleaves in four planes, and sphalerite cleaves in six planes.

Perfect cleavage means that the surface is very smooth; less than perfect cleavage means that the cleavage will be visible but the surfaces have some degree of roughness. The roughness is actually tiny cleavage planes that appear in a stair step pattern with numerous parallel surfaces (**Figure 5.8**); the pattern causes a rough feeling if one runs a thumb along the surface. It's possible to see the tiny stair steps with a hand lens; the stair steps will exhibit the cleavage characteristic of that mineral. Poor cleavage is visible to geologists but rarely to introductory students.

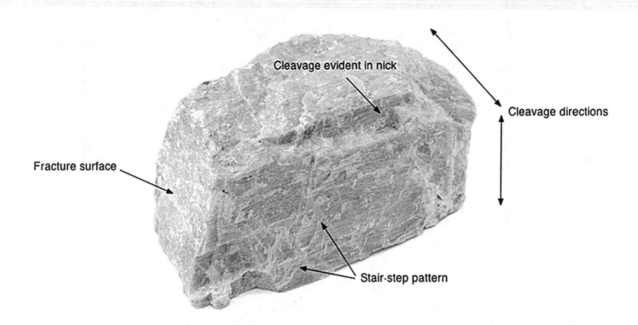

Figure 5.8 Cleavage. Feldspar cleaves in two directions at a 90° angle, and it fractures in the third direction. Note how its cleavage is evident wherever the sample has been nicked or chipped; also note the stair-step pattern in this sample's front surface. Photo: Shutterstock 56068846, credit Tyler Boyes.

Although minerals with cleavage are likely to exhibit cleavage planes, it's unlikely that the entire sample will exhibit one of the shapes seen in **Figure 5.7**. More typically, it is parts of the sample that show the planes and angles described above. For example, in **Figure 5.8**, the feldspar sample shows cleavage in two planes at 90° angles, but that cleavage is clearly evident only in parts of the sample. Besides looking for cleavage in the overall shape of the sample, it may be necessary to examine what appear to be nicks and chips in the sample to determine what, if any cleavage, the sample is exhibiting. Look for partial shapes in the sample similar to those shown in **Figure 5.7** to find areas where the sample's cleavage is evident.

It's often a challenge for students to tell whether the sample they are examining is exhibiting a crystal form or cleavage. After all, crystals and cleavage both present faces. This is especially a challenge for minerals like halite for which the crystal form and cleavage are the same shape. Halite has a cubic crystal structure and cleaves in three planes at 90° to form a cube.

To determine whether a surface is a crystal face or a cleavage plan, examine it closely. A crystal face will be smooth because it grew that way. Cleavage planes are not likely to be perfectly smooth, but have some striations or roughness. Another hint is that crystals often have growth lines that look like repeating patterns of outside edge of the crystal's shape.

Fracture Surfaces

Mineral surfaces that do not cleave will fracture when the mineral breaks. Fractures are produced when there are no planes of weakness or when the chemical bonds holding the mineral structure together extend in all directions, have strong atomic or chemical bonds, and a lack of symmetry of the atomic structure. Even minerals with cleavage may fracture in directions not controlled by the cleavage. Minerals may have cleavage (seen as flat, parallel surface breaks), fracture (seen as curved, irregular surface breaks), or a combination of both. It is not uncommon to see fracture and cleavage surfaces on a single sample. Even so, a sample is said to cleave if it has any cleavage planes.

If a sample does not cleave, all of its surfaces will be fracture surfaces. A fracture surface is uneven, irregular and nonparallel. Quartz is a mineral that does not cleave, but, rather, breaks along smoothly curved surfaces; this type of fracture is called a conchoidal fracture. Conchoidal fractures are often common in glass. Other types of fractures are splintery (like wood) and irregular (like concrete) where cleavage is not present.

Minerals, like feldspar that have two directions of cleavage at right angles also have a fracture direction (see **Figure 5.8**). So as noted above, feldspar samples will usually have some flat surfaces and some ragged edges. The hand lens in your lab kit will be necessary to look closely at the samples for evidence of cleavage and fracture.

Now examine Sample A. Is there evidence of cleavage? _____

If so, what type of cleavage does the sample have? _____

If the sample does not show cleavage, what type of fracture does it have? _____

Specific Gravity

The specific gravity of a mineral describes its density expressed as weight relative to the weight of an equal volume of water. For instance, if a sample weighed exactly what an equal volume of water would weigh, its specific gravity would be 1.0.

Most minerals are substantially heavier than water. For example, the mineral quartz has a specific gravity of 2.65, meaning quartz is 2.65 times heavier than an equal volume of water. Galena, a major source of lead for industry, has a specific gravity of 7.4, making it 7.4 times heavier than an equal volume of water. Gold, with a specific gravity of 19, is 19 times heavier than an equal volume of water.

Accurately determining specific gravity requires specialized equipment (which is not included in your lab kit). In the field, geologists determine relative weight instead by "hefting" the sample. Hefting is done by lifting a piece of one mineral in one hand and lifting an equal-sized piece of another mineral in the other hand. The mineral specimen that feels heavier has a higher specific gravity than one that feels lighter. Heft is described as very light (VL), light (L), medium (M), high (H), or VH (very high). Most minerals samples have a medium heft (which corresponds to a specific gravity of 2–3) so this method is most useful when a sample is exceptionally light or exceptionally heavy.

Heft Sample A several times, comparing it to other mineral samples in your kit. Is Sample A exceptionally light compared to other samples? _____

Is it exceptionally heavy compared to other samples? _____

Would you rate Sample A as VL (very low) L (low), M (medium), H (high), or VH (very high) heft? _____

Additional Tests

The above tests are routinely used on mineral samples to identify them. In addition to the above, there are several tests that are used when a sample is suspected of being a specific mineral. As such, these tests are often conducted after all of the above data has been evaluated.

Acid test. This test is used when a sample is suspected of being calcite or another mineral or rock largely composed of calcium carbonate. Calcite will effervesce (fizz) when treated with dilute hydrochloric acid or with common household acid, like lemon juice or lime juice (citric acid), or vinegar (acetic acid). The fizzing or bubbling action results from the liberation of gaseous carbon dioxide by the reaction of the acid with the calcium carbonate in the mineral. Have patience when attempting tests for this physical property. You may want to scratch your sample, creating a small amount of dust, and then apply the juice or vinegar.

Magnetism. This test is used when a sample is suspected of being magnetite. Some types of magnetite will attract iron or steel objects; if a sample attracts a compass needle or wire nail, it is magnetite. Other types of magnetite do not attract iron or steel objects but will be attracted to a magnet because of its high iron content.

Smell. This test is used when a sample is suspected of being sulfur, kaolinite or another mineral with a distinctive smell. A sample containing sulfur will smell like rotten eggs or a match just after having been struck. Kaolinite has a distinctive musty smell.

Taste. This test is used only when a sample is suspected of being halite, kaolinite, or another mineral with a distinctive taste. Halite will have a familiar salty taste and kaolinite will taste musty. Do not use this test routinely on every sample, as there are minerals that contain toxic elements, like mercury, arsenic and lead!

Using the Mineral Identification Chart

Once the relevant tests have been conducted, the final step in mineral identification is to compare the data that has been gathered to a mineral identification chart or key. The Mineral Identification Key that appears in Appendix V of this book is an adaptation of one provided by the United States Geological Service. It lists common minerals in two tables: one for minerals with a metallic luster and one for minerals with a nonmetallic luster. Within these two tables, minerals are presented in order of their hardness. The chart provides color, streak, cleavage, and miscellaneous other properties as well.

Enter the data you've gathered about Sample A to the Student Chart in **Figure 5.9**. Then compare this data to the data in the Mineral Identification Key. If the chart points you to a mineral that could be positively identified with one of the miscellaneous tests described above, conduct that test. Then arrive at a conclusion about the identity of Sample A.

What mineral is Sample A? _____

Lab Exercise

Lab Exercise: Mineral Identification

In this laboratory exercise, you will identify the minerals in your lab kit. The ability to identify minerals is one of the most fundamental geological skills. It is an essential skill in identifying rocks, for one must first identify the minerals within a rock before the rock itself can be identified. Only after minerals and rocks have been identified can their origin, classification, and alteration be adequately understood. The ability for a geologist to identify minerals by using the simplest tools is often a necessity, particularly when working in the field.

Instructions and Observations

Remove Samples B through I from the bag marked Lab #5 Mineral Samples in your lab kit. Use the steps indicated in the flow chart in **Figure 5.10** to make your identification. Use the chart in **Figure 5.9** below to keep track of each sample's properties. You should have already filled in the properties and the mineral name for Sample A.

When you're finished, submit your results to your instructor as directed. Keep the completed chart to answer quiz questions about the minerals you have identified.

Sample	Color	Luster	Streak	Hardness	Crystal Form	Cleavage	Specific Gravity	Misc. Tests	Mineral Name
A									
B									
C									
D									
E									
F									
G									
H									
I									

Figure 5.9 Student Chart.

Mineral Identification Step-by-Step

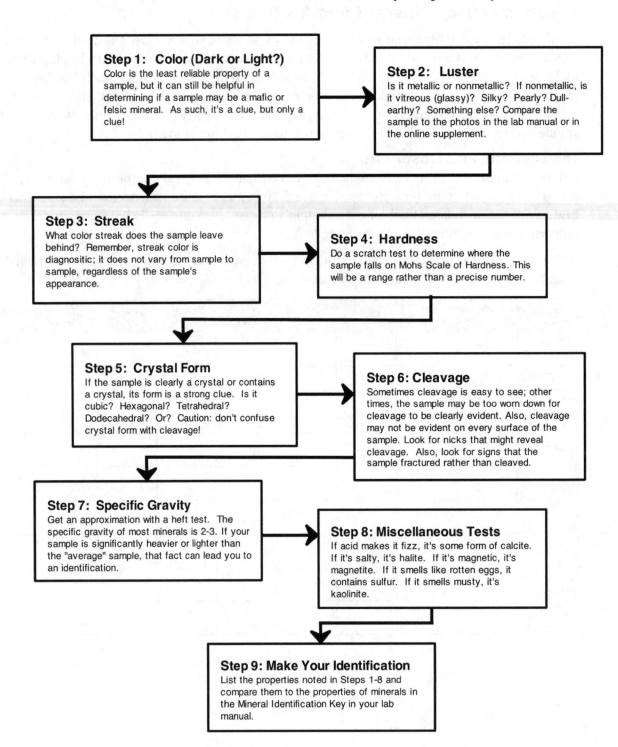

Step 1: Color (Dark or Light?)
Color is the least reliable property of a sample, but it can still be helpful in determining if a sample may be a mafic or felsic mineral. As such, it's a clue, but only a clue!

Step 2: Luster
Is it metallic or nonmetallic? If nonmetallic, is it vitreous (glassy)? Silky? Pearly? Dull-earthy? Something else? Compare the sample to the photos in the lab manual or in the online supplement.

Step 3: Streak
What color streak does the sample leave behind? Remember, streak color is diagnositic; it does not vary from sample to sample, regardless of the sample's appearance.

Step 4: Hardness
Do a scratch test to determine where the sample falls on Mohs Scale of Hardness. This will be a range rather than a precise number.

Step 5: Crystal Form
If the sample is clearly a crystal or contains a crystal, its form is a strong clue. Is it cubic? Hexagonal? Tetrahedral? Dodecahedral? Or? Caution: don't confuse crystal form with cleavage!

Step 6: Cleavage
Sometimes cleavage is easy to see; other times, the sample may be too worn down for cleavage to be clearly evident. Also, cleavage may not be evident on every surface of the sample. Look for nicks that might reveal cleavage. Also, look for signs that the sample fractured rather than cleaved.

Step 7: Specific Gravity
Get an approximation with a heft test. The specific gravity of most minerals is 2-3. If your sample is significantly heavier or lighter than the "average" sample, that fact can lead you to an identification.

Step 8: Miscellaneous Tests
If acid makes it fizz, it's some form of calcite. If it's salty, it's halite. If it's magnetic, it's magnetite. If it smells like rotten eggs, it contains sulfur. If it smells musty, it's kaolinite.

Step 9: Make Your Identification
List the properties noted in Steps 1-8 and compare them to the properties of minerals in the Mineral Identification Key in your lab manual.

Figure 5.10 Mineral Identification Flow Chart. Illustration by Susan Wilcox.

Online Activities

Per your instructor's directions, go to the online supplement for this lab and complete the activities assigned. Viewing the online videos will help you to complete the quiz.

Quiz

Multiple Choice

Questions 1 through 21 are based on the **Lab Exercise:** *Mineral Identification.*

1. What mineral is specimen A?
   a. gypsum
   b. calcite
   c. quartz
   d. feldspar

2. Specimen B has
   a. a dull or earthy luster.
   b. a silky luster.
   c. a metallic luster.
   d. a glassy luster.

3. The hardness of Specimen B is
   a. < 2.5
   b. 2.5–3.5
   c. 3.5–4.5
   d. 4.5–5.5
   e. > 5.5

4. Specimen B is
   a. fluorite.
   b. calcite.
   c. quartz.
   d. feldspar.

5. Specimen C has
   a. cleavage in one plane.
   b. cleavage in two planes at 90° angles.
   c. cleave in three planes not at 90° angles.
   d. cleavage in three planes at 90° angles.

6. Specimen C is
   a. hornblende.
   b. magnetite.
   c. galena.
   d. biotite.

7. Specimen D has
   a. cleavage in two planes not at 90° angles.
   b. cleavage in three planes not at 90° angles.
   c. cleavage in three planes at 90° angles.
   d. conchoidal fracture.

8. Specimen D
   a. effervesces.
   b. attracts a magnet.
   c. feels greasy.
   d. has a very high specific gravity.

9. Specimen D is
   a. fluorite.
   b. gypsum.
   c. calcite.
   d. feldspar

10. The hardness of Specimen E is
    a. < 2.5.
    b. 2.5–3.5.
    c. 3.5–4.5.
    d. 4.5–5.5.

11. Specimen E is
    a. fluorite.
    b. gypsum.
    c. calcite.
    d. quartz.

12. Specimen F shows cleavage in
    a. 1 plane.
    b. 2 planes at 90° angles.
    c. 3 planes not at 90° angles.
    d. 4 planes.

13. Specimen F is
    a. magnetite.
    b. fluorite.
    c. halite.
    d. biotite.

14. Specimen G
    a. effervesces.
    b. attracts a magnet.
    c. has six cleavage directions.
    d. feels greasy.

15. Specimen G is
    a. hornblende.
    b. magnetite.
    c. pyrite.
    d. galena.

16. Specimen H has
    a. cleavage in one plane.
    b. cleavage in two planes not at 90° angles.
    c. cleavage in three planes at 90° angles.
    d. cleavage in three planes not at 90° angles.

17. Specimen H is
    a. magnetite.
    b. muscovite.
    c. halite.
    d. biotite.

18. Examine the cleavage of specimen I. Specimen I has
    a. cleavage in one plane.
    b. cleavage in two planes at 90° angles.
    c. cleavage in two planes not at 90° angles.
    d. no cleavage.

19. Specimen I is
    a. hornblende.
    b. fluorite.
    c. halite.
    d. biotite.

20. Specimen J has
    a. cleavage in one plane.
    b. cleavage in two planes not at 90° angles.
    c. cleavage in three planes not at 90° angles.
    d. cleavage in three planes at 90° angles.

21. Specimen J is
    a. magnetite.
    b. biotite.
    c. muscovite.
    d. pyrite.

22. When a mineral breaks, it often does so along its plane or planes, which is known as
    a. luster.
    b. streak.
    c. hardness.
    d. cleavage.

23. What percentage of Earth's crust is composed of the mineral group, silicates?
    a. 60%
    b. 70%
    c. 80%
    d. 90%

24. An example of a mineral composed of isolated tetrahedra is the gemstone peridot. Peridot is a member of the _____ group, the largest group of minerals with isolated tetrahedron structure.
    a. amphibole
    b. mica
    c. olivine
    d. pyroxene

25. The mineral structures composed of isolated tetrahedra have no planes of weakness, which means minerals such as garnet have no
    a. cleavage.
    b. luster.
    c. streak.
    d. color.

26. Which mineral group is formed of double chains of silicon-oxygen tetrahedra?
    a. the olivine group
    b. the pyroxene group
    c. the amphibole group
    d. the mica group

27. Which mineral group has silicon-oxygen tetrahedra bonded in a sheet structure?
    a. the olivine group
    b. the pyroxene group
    c. the amphibole group
    d. the mica group

28. Quartz and feldspar share a similar _____ silicate structure.
    a. single-chain
    b. double-chain
    c. sheet
    d. framework

29. _____ minerals are dark-colored because they contain _____.
    a. Mafic; iron and/or magnesium
    b. Mafic; calcium, sodium, or potassium
    c. Felsic; calcium, sodium, or potassium
    d. Felsic; iron and/or magnesium

Short Answer

1. What is the difference between a rock and a mineral?

2. Where might you find a use of gypsum in your home?

3. Name at least two uses for halite.

4. What are three uses for quartz?

5. What is a mafic mineral and which mineral groups are considered mafic?

6. What is a felsic mineral and which mineral groups are considered felsic?

7. Explain how to determine the hardness of a mineral.

Student Answer Key

Page 108:

Take a look at Sample A and note its color.

It is pink.

Page 109:

What type of luster does Sample A appear to have?

It has pearly luster.

Page 110:

Can you see a streak?

No, it is too light to see.

Page 111:

What hardness range does it have on Mohs scale?

a. 1–2.5

b. 2.5–3.5

c. 3.5–5.5

d. > 5.5

Page 112:

Examine Sample A with and without the magnifying glass contained in your lab kit. Can you see any crystals?

Yes.

If so, does it have one of the crystal forms described above?

No.

Page 115:

Now examine Sample A. Is there evidence of cleavage?

Yes.

If so, what type of cleavage does the sample have?

Two planes at 90° angles.

If the sample does not show cleavage, what type of fracture does it have?

Sample A shows cleavage planes.

Page 116:

Is Sample A exceptionally light compared to other samples?

No.

Is it exceptionally heavy compared to other samples?

No.

Would you rate Sample A as VL (very low) L (low), M (medium), H (high), or VH (very high) heft?

M

What mineral is Sample A?

It is feldspar.

IGNEOUS ROCKS AND VOLCANISM

Lesson 6

Purpose

In this laboratory lesson, you will identify different types of igneous rocks by learning the origins of how these different types of igneous rocks formed through volcanic activity. Nearly 80 percent of Earth's surface, both above and below sea level, contains igneous rocks that were formed by volcanism.

Learning Objectives

After completing this laboratory lesson, you will be able to:

- Explain how igneous rocks form and the difference between extrusive and intrusive igneous rocks.
- Explain how the environment in which magma cools affects the texture of the resulting igneous rocks.
- Describe how magma composition determines the style of a volcanic eruption and the type of volcanic structure that will be formed.
- Identify different types of igneous rocks based on their texture and composition (i.e., color, mineralogy, and so forth).

Materials Needed

- ❏ 10X-magnifying hand lens (in lab kit)
- ❏ Ten igneous rock specimens (in lab kit)
- ❏ Pencil
- ❏ Eraser
- ❏ Metric ruler (in lab kit)
- ❏ White sheet of paper

Overview

Earth is a planet of constant change, although many changes are so slow they are imperceptible to us. Among them are the changes rocks undergo. While rocks may seem permanent to us, most rocks will go through many types of change over the course of their existence.

There are three main types of rocks: igneous rocks, sedimentary rocks, and metamorphic rocks. Sedimentary and metamorphic rocks are the subjects of Lessons 7 and 8, respectively. The rock cycle seen in **Figure 6.1** below illustrates how rocks in each category form, and the changes each type of rock may undergo to become a different type of rock.

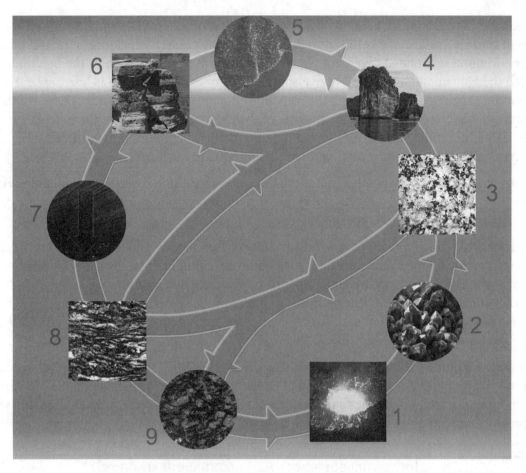

Figure 6.1 The Rock Cycle. Illustration by Rick de Goede.

While the rock cycle is ongoing, discussion of it usually begins with molten rock, magma (1). When magma (1) cools, it crystallizes (2) into igneous rock (3), the topic of this lesson. Wind, water and ice weather rocks(4) into particles or their component chemicals, which may then be deposited in layers long enough to turn into (5) sedimentary rock (6). Or rocks may be carried deep underground (7) by tectonic processes where conditions are extreme enough to transform them into metamorphic rock (8). If carried deeply enough underground, temperatures and pressures may be high enough to melt the rock (9) once again into magma (1).

Note that rocks don't necessarily go through these steps in order; the conditions to which it is exposed determines what the next step for a particular rock will be.

Igneous Rock Formation

Because certain types of changes produce specific types of rocks, to a geologist, rocks are more than a collection of minerals. Every rock tells a story that provides insight into Earth's history in the region where the rock formed or where it was found. Igneous rocks are those that solidified directly from magma or lava; therefore, the story of any igneous rock is one that includes the melting of rock, its rise through the mantle and crust, and, finally, its solidification far underground or its dramatic extrusion at the surface as the result of volcanic activity. Geologists have learned to read these rocks to unravel their stories, each one of which is a clue to the larger story of Earth's history.

The story of an igneous rock begins with the formation of magma. Magma is molten rock beneath Earth's surface, the result of rock melting as the result of heat, pressure and other factors. Magma is found at divergent plate boundaries, at hot spots such as the one that created the Hawaiian Islands, and above subduction zones. The magma is less dense than surrounding rock so it rises through the rock until it approaches the surface. There it forms a magma chamber. In some cases, the magma in the chamber solidifies in place. In other cases, a conduit to the surface forms, and the magma flows through it to extrude at the surface as lava in a volcanic eruption; there it subsequently solidifies.

Igneous rocks that form when magma solidifies beneath Earth's surface are called *intrusive* or *plutonic* igneous rocks, whereas those that formed by cooling at Earth's surface are called *extrusive* or *volcanic* igneous rocks.

The process by which molten rock solidifies to become igneous rock is called crystallization because it is during the change of state from liquid to solid that the individual mineral crystals form. Igneous rocks are all composed of minerals belonging to the silicates group; the exact silicate minerals appearing in an individual rock sample is determined partly by the composition of the molten rock from which it formed and partly by the temperature at which the minerals crystallized.

The chart in **Figure 6.2** is called Bowen's reaction series. It illustrates the order in which the various silicate minerals solidify as magma cools; the chart also explains one reason why certain minerals are likely to be found together in igneous rocks (which helps in identifying the rocks). Rocks that solidify at the same time are more likely to be found together.

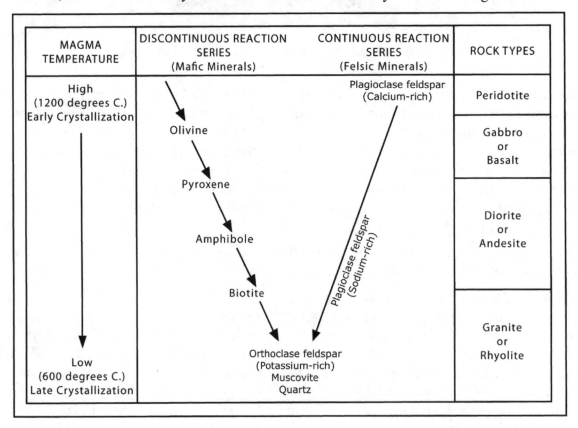

Figure 6.2 Bowen's Reaction Series. Illustration by Mark Worden.

The chart includes two series of minerals: the discontinuous series and the continuous series. The discontinuous series (on the left side of the chart) show that as molten rock cools, the first minerals to solidify are those with isolated tetrahedra, the olivine group. As the cooling continues, the tetrahedra begin to link together to form crystals with a single chain structure, producing the pyroxene group. Mineral crystals with a double-chain structure form next, producing the amphibole group (like the hornblende from your Lesson 5 lab kit). As temperatures further decrease, the tetrahedra link together in a sheet structure, forming the mica group. This series is called discontinuous because as magma cools, the new minerals react with it to begin forming the crystalline structures of the next group.

The right-hand, continuous series side of the chart shows the order of crystallization for silicates with framework structures. Calcium-rich plagioclase feldspar forms first, then sodium-rich plagioclase, and finally, at the cooler temperatures, potassium-rich orthoclase. All of these are feldspars but differ in the relative amounts of calcium, sodium, or potassium they contain within their structures, which affect their crystallization temperature. At the very bottom of the chart is quartz, the last silicate to crystallize as magma cools.

Bowen's reaction series can be used in reverse to predict the order in which solid minerals melt. As temperatures and pressures rise, minerals at the bottom of the chart will melt first, followed by those in the middle, and, finally, those at the top of the chart. This sequence helps to explain why not all magma has the same composition. If the temperature and pressure are high enough to melt quartz, orthoclase, and mica but not pyroxenes, olivine and plagioclase, the resulting magma will be composed mostly of the elements found in felsic minerals. Such magmas are referred to as felsic magmas. By the same token, if temperatures and pressures are high enough for pyroxenes, calcium-rich plagioclase, and olivine to melt, the magma will be dominated by the elements found in mafic minerals and is known as a mafic magma.

As might be expected, different magma compositions will result in different igneous rock compositions as the magma cools down enough for the minerals to once again solidify.

Looking at the chart, then, it's no surprise that most granite includes feldspar, quartz and biotite. First, they are composed of the elements found in a felsic magma. Second, these minerals crystallize at about the same, cooler temperatures. Igneous rocks, like granite, that contain large amounts of felsic minerals are referred to as felsic igneous rocks. Igneous rocks containing large amounts of olivine, pyroxene, and calcium-rich plagioclase are referred to as mafic rocks; even though they contain plagioclase, which is felsic, the large amounts of mafic minerals give these rocks the same dark color associated with mafic minerals. Rocks that contain balanced mixtures of mafic and felsic minerals are said to have intermediate composition.

The Volcanic Origins of Extrusive Igneous Rocks

An erupting volcano is one of the most impressive, and among the most dangerous, of geological events. Some volcanoes erupt with a brilliant display of glowing lava running down their flanks; others erupt with explosions that send ash and rocks, called pyroclastic materials, high into the atmosphere. The type of eruption that occurs is not a matter of chance; it is the result of the composition of the magma associated with the volcano.

Very hot, fluid magmas are mafic in composition; recall the high temperatures at which mafic minerals melt. These magmas as associated with divergent plate boundaries and many hot spots, such as the one that has formed the Hawaiian Islands. Volcanic eruptions involving mafic magmas are relatively quiet as the lava extrudes from the crater of the volcano or from fissures along its flanks, and flows downhill until it cools enough to solidify. The volcanic structure formed from this type of eruption is called a shield volcano, characterized by long, gradual slopes (**Figure 6.3**). The younger volcanoes of the Big Island of Hawaii are good examples of shield volcanoes.

Figure 6.3 Shield Volcano vs. Stratovolcano. Photos: (left) Shutterstock 18745270, credit Jennie Endriss; (right) Shutterstock 2456346, credit TTPhoto.

This type of lava may also flow from long fissures in the continental crust; these fissure eruptions don't build a volcanic structure but flood the nearby region to create lava plateaus. Because mafic lava solidifies into the mafic rock, basalt (the same rock that composes the oceanic crust), these rocks are often called flood basalts.

Felsic magmas are cooler than mafic magmas and, due to a higher silica content, are far more viscous. Viscosity is defined as resistance to flow; consider the manner in which the more viscous maple syrup flows compared to less viscous water. A viscous magma does not rise to the surface as easily as a fluid magma, and it is likely to trap the gases contained in magma. As the magma rises, the trapped gas bubbles expand in volume; if they continue to be trapped, eventually the gases will escape with an explosion that sends pyroclastic materials—particles of solidified lava ranging in size from less than 2 mm (ash) to many feet across (volcanic bricks and bombs)—high into the atmosphere. Once the supply of gases and pyroclastic materials have been depleted, felsic lava begins to extrude. The lava is very thick and does not flow far (compared to mafic lava) before it solidifies.

Such explosive eruptions are characteristic of felsic and intermediate magmas and are associated with the volcanic arcs that form near convergent boundaries over subduction zones. The 1982 eruptions of Mount St. Helens in the state of Washington and the 1991 eruption of Mount Pinatubo in the Philippines are examples of this type of explosive eruption due to the more viscous magma.

The shape of volcanoes formed by felsic and intermediate eruptions is steeper than that of shield volcanoes. Each eruption deposits a layer of pyroclastic materials followed by thick felsic or intermediate magma. Over time, successive eruptions form a cone-shaped structure called a stratovolcano (**Figure 6.3**). Japan's Mount Fuji is perhaps the world's most recognizable example of a stratovolcano.

Another structure formed by pyroclastic materials is the cinder cone. As lava, either mafic or felsic, fountains into the air, the blobs may solidify into small rocks, or cinder, containing many air bubbles. Over time, the cinders accumulate around the vent into a cone shape. Wizard's Island in Crater Lake, Oregon, is one example of a cinder cone.

Igneous Rock Textures

A geologist can often tell by an igneous rock's appearance whether an igneous rock formed underground, at the surface, or in midair. Igneous rocks that solidify underground differ in appearance than those that form at Earth's surface. Pyroclastic materials that solidified in mid-air have their own characteristic appearance. Texture is the term geologists use to refer appearance of a rock as determined by the size, shape, and arrangement of its grains, as the crystals in igneous rocks are often referred to. There are several types of textures that result from the conditions under which igneous rocks form.

Phaneritic Texture (coarse-grained)

The size of the grains in an igneous rock generally indicates the rate at which the magma or lava cooled and crystallized. For example, mineral crystals that form in a slowly cooling magma within the earth's crust (intrusive igneous rocks) become large because they have a long time to grow before the magma completely solidifies. This large-grained, or coarse-grained texture, is referred to as phaneritic (**Figure 6.4**). A rock has a phaneritic texture if it has uniformly sized crystals that are visible to the naked eye. The dimensions of the individual mineral grains range from about 1 mm to about 10 mm (1 cm).

Figure 6.4 Phaneritic vs. Aphanitic Texture. Photos: (left) Shutterstock 13546804, credit Maxim Tupikov; (right) Shutterstock 50419654, credit Schank.

Aphanitic Texture (fine-grained)

Mineral grains that form in a rapidly cooling lava on the earth's surface (extrusive igneous rocks) will be much smaller, often microscopic, because the crystals have relatively little time to grow. This texture is called aphanitic (**Figure 6.4**). A texture is aphanitic if the grains can't be seen with the naked eye; the dimensions of aphanitic grains are less than 1 mm.

Porphyritic Texture (mixed grain)

A mixed texture results when magma cools in a two-stage process: it first cools slowly beneath the surface, during which large crystals, called phenocrysts, form; then conditions change that make it cool more quickly. As a result, the rock surrounding the phenocrysts, called the groundmass, is finer grained (**Figure 6.5**). The groundmass may be either phaneritic or aphanitic. Many types of rocks can take a porphyritic texture if subjected to this two-step cooling process. For instance, granite with phenocrysts would be called porphyritic granite; andesite containing phenocrysts would be called porphyritic andesite.

Figure 6.5 Porphyritic Texture. Photo: © Albert Copley/Visuals Unlimited, Inc.

Pegmatitic Texture (large crystals)

If the mineral crystals grow to be very large, more than 1 cm across, the rock is referred to as a pegmatite. Pegmatites are often impressive in their beauty and uniqueness, and so they are favored by rock collectors (**Figure 6.6**).

Figure 6.6 Pegmatitic Texture. Photo: Shutterstock 6668629, credit Jens Mayer.

Vesicular Texture

Some extrusive rocks cool so quickly that volcanic gas bubbles are trapped inside them. This texture is called vesicular, referring to the vesicles, or holes, made by the gas bubbles (**Figure 6.7**). There are two major types of vesicular igneous rocks: pumice and scoria. Scoria is porous, dark, igneous rock that contains so many vesicles that it resembles a sponge. Pumice is a light-colored, frothy glass with so many tiny vesicles and such a low density that it floats on water.

Figure 6.7 Vesicular Texture. Scoria (left) and pumice (right) have vesicular textures formed either when they solidified midair after being ejected from a volcano or from lava as gas bubbles escaped. Pumice is light enough to float on water. Photos: Shutterstock 56068849 (left) and 56068843 (right), credit Tyler Boyes.

Some igneous rocks form when lava cools in midair after being thrown high into the atmosphere by an explosive volcanic eruption. Materials ejected during such an eruption are said to be pyroclastic, and most of them are vesicular. This solidified volcanic debris is called tephra, and it is classified by its shape and size:

- Volcanic ash is less than 2 mm in diameter.
- Lapilli, or cinders, range from 2 mm to 64 mm in diameter (**Figure 6.8**).
- Volcanic bombs and blocks are 64 mm and larger in diameter.

Figure 6.8 Cinder. Cinders are vesicular rocks between 2 and 64 millimeters in diameter that form when lava solidifies in midair. Photo: Shutterstock 26586715, credit Only Fabrizio.

Glassy Texture
If lava is cooled or quenched very suddenly, there is no time for mineral grains to grow, and an amorphous volcanic glass called obsidian forms instead (**Figure 6.9**). Obsidian has no grain and resembles a chunk of glass. Obsidian's composition makes it a felsic rock; its dark color is the result of iron and magnesium impurities in the felsic magma.

Figure 6.9 Glassy Texture. Photo: Shutterstock 56061253, credit Tyler Boyes.

Pyroclastic Texture

Some igneous rocks are composed of pyroclastic materials welded together. Tuff is a pyroclastic rock largely composed of volcanic ash surrounding fragments of pyroclastic rocks and other volcanic debris. (**Figure 6.10**). Volcanic breccia is composed chiefly of cinders, volcanic bombs, and volcanic blocks that have been cemented together. Volcanic breccia thus contains fairly large particles, whereas tuff contains tinier particles. Tuff often contains phenocrysts as well. The phenocrysts are recognizable by their glassy luster and sharp edges, whereas the fragments in tuff are generally dull and have irregular shapes.

Figure 6.10 Pyroclastic Texture. Tuff is an igneous rock with a pyroclastic texture. It forms from an accumulation of volcanic ash that cements together, and it typically includes pyroclastic rocks and debris fragments. Photo: Courtesy of Mark A. Wilson.

Igneous Rock Identification

There are many different types of igneous rocks, but some of the most common appear in **Figure 6.11** below. Follow along on this chart as you note how the rocks identification is based on both composition and texture.

	Aphanitic	Phaneritic
Ultramafic	Komatiite	Peridotite
Mafic	Basalt	Gabbro
Intermediate	Andesite	Diorite
Felsic	Rhyolite	Granite

Figure 6.11 Igneous Rock Classification Chart. Photos: Komatiite, Released into the public domain by the photographer via Wikipedia; Peridotite, Courtesy of Rick de Goede; Basalt, Shutterstock 56068825, credit Tyler Boyes; Gabbro, Shutterstock 56068804, credit Tyler Boyes; Andesite, Shutterstock 56068873, credit Tyler Boyes; Diorite, Shutterstock 56068861, credit Tyler Boyes; Rhyolite, Shutterstock 56068864, credit Tyler Boyes; Granite, Shutterstock 56061274, credit Tyler Boyes.

The most abundant igneous rock, because it comprises the oceanic crust, is basalt. Basalt is a mafic rock with an aphanitic texture; it solidifies from the very hot, fluid mafic magma that exudes at rift zones and hot spots. This magma and the resulting rock are rich in olivine, pyroxene and calcium-rich plagioclase—just what would be expected, based on Bowen's reaction series. This produces a very dark brown or black rock. If the same magma solidifies underground, its texture will be phaneritic; in this case, the resulting rock is called gabbro. Note that basalt and gabbro are similar in composition; the difference between them is where they crystallized and the resulting texture.

The most common continental igneous rock is granite. Granite is a felsic rock with a phaneritic texture; it solidified from a cooler, thicker felsic magma that didn't make it to Earth's surface but crystallized beneath the surface. Granite is rich in quartz, orthoclase, biotite and often contains some plagioclase and hornblende; either combination gives granite a light color with dark flecks (from the biotite and/or hornblende). California's Sierra Nevada mountain range is a granitic mass that became exposed at the surface long after crystallization.

When a felsic magma does break through the surface, it will cool quickly and have an aphanitic texture; in this case, the resulting rock is called rhyolite. Rhyolite is more uncommon than granite (for the reasons just cited); however, there are significant deposits in the Yellowstone National Park area and across the western United States as rhyolite is associated with continental hot spots.

Intermediate magma has a balance of mafic and felsic minerals and produces rocks said to have intermediate composition. When intermediate magma crystallizes beneath the surface, it forms the phaneritic intermediate rock, diorite. Diorite is distinguished from granite by a greater abundance of mafic minerals. It often has a salt-and-pepper appearance.

If an intermediate magma reaches the surface, rapid cooling will produce the aphanitic intermediate rock, andesite. Andesite is associated with continental arcs at convergent plate boundaries, like South America's Andes Mountains for which this rock is named.

Any of these rocks can form in the two-step cooling process that produces a porphyry. When they do, the adjective "porphyritic" is used to describe them. Rhyolite and andesite are frequently porphyritic, but there is also porphyritic granite and porphyritic basalt. Identifying a porphyritic rock requires the same process as identifying a nonporphyritic rock; you would identify it as having a porphyritic texture, then go on to identify the rock that composes its groundmass. The andesite and rhyolite that appear in **Figure 6.11** are both actually porphyritic with an aphanitic groundmass.

A less common type of magma contains an unusually large amount of ferromagnesian minerals, and so is called ultramafic. This type of magma crystallizes to form ultramafic igneous rocks. Peridotite is a phaneritic ultramafic rock which is composed of olivine, pyroxene, and/or calcium-plagioclase. Peridotite composes most of Earth's mantle; its aphanitic equivalent, komatiite, is very rare and usually very old as most ultramafic magmas solidify before reaching Earth's surface.

Vesicular rocks are recognizable by their vesicles and may have a felsic, intermediate or mafic composition Pyroclastic rocks include vesicular rocks, like scoria, but the term, pyroclastic texture, refers to rocks that are solid and contain fragments of volcanic debris. These rocks are typically felsic because pyroclastic materials are associated with the explosive eruptions caused by felsic magmas. The pyroclastic rock tuff has the finest texture because it is composed of volcanic ash welded together; however, tuff often has other visible pyroclastic fragments as well. This gives it a look similar to that of porphyritic rocks; the difference is that the phenocrysts in porphyritic rocks are crystals and the fragments in pyroclastic rock are volcanic debris. As with porphyry, tuff may carry a label that indicates its composition: rhyolite tuff, for instance.

Lab Exercises

Lab Exercise #1: *Identification of Igneous Rocks*

The identification and classification of igneous rocks is based on texture and mineralogical composition. Color is also useful because it is a reflection of the mineralogical composition, but, if you can see the grain, you may also be able identify the mineral based on cleavage or a crystal form characteristic of a component mineral. Feel free to use the Mineral Identification Key in Appendix V of this lab manual to identify component minerals, or to review the material in Lesson 5, as needed.

In addition, you may download and print the Igneous Rock Textures Photo Guide to help you to identify the samples in this lesson.

Instructions and Observations

Step 1: Retrieve the bag labeled Lab #6 Igneous Rock Samples from your lab kit and place the specimens (numbered 1 through 10) on a white sheet of paper.

Step 2: Determine the identity of each rock, then write its name in the proper cell of **Figure 6.12**. Note that not every cell will be filled. The flow chart in **Figure 6.13** will help you narrow down the possibilities.

Texture	Composition		
	Felsic	Intermediate	Mafic
Aphanitic			
Phaneritic			
Porphyritic			
Pegmatitic			
Vesicular			
Pyroclastic			
Glassy			

Figure 6.12 Igneous Rock Lab Chart.

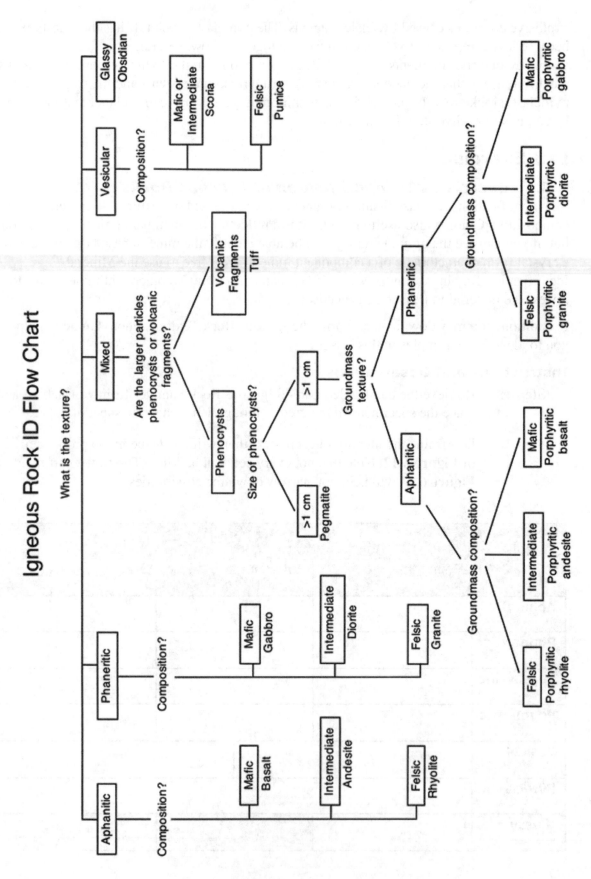

Figure 6.13 Igneous Rock Identification Flow Chart. Illustration by Susan Wilcox.

Lab Exercise #2: *Magma, Volcanism, and Igneous Rocks*

Every igneous rock has a story to tell that geologists can unravel by considering the rock's texture and composition. In this exercise, you will briefly tell the story of four of the igneous samples you just identified—samples 1, 4, 6, and 9.

Instructions and Observations

Write a few lines about each of the rocks listed below, including the following information:

(1) Rock name

(2) Was this rock formed from felsic, intermediate, or mafic magma or lava? How do you know?

(3) What type of plate boundary is associated with this type of magma/lava composition?

(4) What kind of eruption and volcanic structure is this type of magma/lava associated with?

(5) Where did this rock solidify? Underground? On Earth's surface from extruded lava? In midair as part of expelled pyroclastic material? How do you know?

(6) How relatively quick was this rock's formation? Did it solidify in a one- or two-step process? How do you know?

Submit to your instructor as directed. Keep a copy for yourself to use in answering quiz questions.

Sample #1:

Sample #4:

Sample #6:

Sample #9:

Online Activities

Per your instructor's directions, go to the online supplement for this lab and complete the activities assigned. Viewing the online videos will help you to complete the quiz.

Quiz

Multiple Choice

1. What mineral composition is most characteristic of felsic rocks?
   a. olivine, pyroxene, and calcium-rich plagioclase
   b. orthoclase, quartz, and biotite
   c. calcium-rich plagioclase and hornblende with some olivine
   d. particles of volcanic ash welded together

2. What mineral composition is most characteristic of mafic rocks?
   a. olivine, pyroxene, and calcium-rich plagioclase
   b. orthoclase, quartz, and biotite
   c. sodium-rich plagioclase, hornblende, and quartz
   d. particles of volcanic ash welded together

3. What mineral composition is most characteristic of intermediate rocks?
   a. olivine, pyroxene, and calcium-rich plagioclase
   b. orthoclase, quartz, and biotite
   c. sodium-rich plagioclase and hornblende with some quartz or biotite
   d. particles of volcanic ash welded together

4. When mafic lava breaks through fissures in the continental crust, which of the following often results?
   a. lava plateaus
   b. composite volcanoes
   c. lava domes
   d. pyroclastic sheets

5. Which scenario is associated with mafic magma?
   a. hot, fluid lava that forms a shield volcano
   b. hot, fluid lava that forms a stratovolcano
   c. an explosive eruption that ejects pyroclastic materials
   d. cool, viscous magma that forms a shield volcano

6. Which scenario is associated with felsic magma?
   a. hot, fluid lava that forms a shield volcano
   b. basaltic lava that forms a stratovolcano
   c. an explosive eruption that ejects pyroclastic materials
   d. cool, viscous magma that forms a shield volcano

7. Wizard Island in Crater Lake, Oregon, is an example of a
   a. cinder cone.
   b. volcanic neck.
   c. lava dome.
   d. composite volcano.

8. A porphyritic rock forms when
   a. blobs of lava solidify in midair.
   b. there is a period of slow cooling, followed by more rapid cooling.
   c. magma cools slowly underground until it completely solidifies.
   d. a lava flow cools rapidly at the surface.

9. A porphyritic rock contains both _____ and _____.
   a. vesicles; crystals
   b. phenocrysts; groundmass
   c. pegmatites; groundmass
   d. phenocrysts; vesicles

Questions 10 through 25 are based on **Lab Exercise #1:** *Identification of Igneous Rocks.*

10. Specimen #2 has a
    a. pegmatitic texture.
    b. phaneritic texture.
    c. aphanitic texture.
    d. glassy texture.

11. Specimen #2 crystallized from
    a. a felsic magma.
    b. an intermediate magma.
    c. a mafic magma.
    d. an ultramafic magma.

12. Specimen #2 is
    a. gabbro.
    b. diorite.
    c. syenite.
    d. granite.

13. Specimen #5 is
    a. andesite.
    b. basalt.
    c. pumice.
    d. obsidian.

14. Specimen #6 is
    a. andesite.
    b. basalt.
    c. pumice.
    d. obsidian.

15. Specimen #7 has
    a. a pegmatitic texture.
    b. a phaneritic texture.
    c. an aphanitic texture.
    d. a glassy texture.

16. Specimen #7 crystallized from
    a. a felsic magma.
    b. an intermediate magma.
    c. a mafic magma.
    d. an ultramafic magma.

17. Specimen #7 is
    a. gabbro.
    b. diorite.
    c. gray granite.
    d. red granite.

18. Specimen #8 is
    a. an ultramafic igneous rock.
    b. a mafic igneous rock.
    c. an intermediate igneous rock.
    d. a felsic igneous rock.

19. Specimen #8 is
    a. rhyolite.
    b. porphyritic andesite.
    c. basalt.
    d. pumice.

20. Specimen #9 crystallized from
    a. a felsic magma.
    b. an intermediate magma.
    c. a mafic magma.
    d. a mafic lava.

21. Specimen #9 is
    a. gabbro.
    b. diorite.
    c. gray granite.
    d. red granite.

22. Specimen #10 is
    a. rhyolite.
    b. andesite.
    c. basalt.
    d. obsidian.

23. Specimen #1 is
    a. tuff.
    b. basalt.
    c. pumice.
    d. granite.

24. Specimen #3 is
    a. tuff.
    b. andesite.
    c. red granite.
    d. obsidian.

25. Specimen #4 is
    a. gabbro.
    b. basalt.
    c. rhyolite.
    d. obsidian.

Questions 26 through 30 are based on **Lab Exercise #2:** *Magma, Volcanism, and Igneous Rocks.*

26. Which of the following statements best describes Sample #1?
    a. It formed slowly underground from felsic magma.
    b. It formed quickly in mid-air from felsic lava.
    c. It formed quickly on the surface from mafic lava.
    d. It formed slowly underground from mafic magma.

27. Which of the following statements best describes Sample #4?
    a. It formed slowly underground from felsic magma.
    b. It formed quickly in mid-air from mafic lava.
    c. It formed quickly on the surface from felsic lava.
    d. It formed slowly underground from mafic magma.

Lesson 6/Igneous Rocks and Volcanism   **145**

28. What location is associated with the composition of the magma that formed Sample #4?
    a. continent-continent convergent plate boundary.
    b. continental hot spots.
    c. transform boundary
    d. divergent boundary

29. Which of the following statements best describes Sample #6?
    a. It formed slowly underground from felsic magma.
    b. It formed from felsic lava as gas bubbles escaped.
    c. It formed quickly on the surface from mafic lava.
    d. It formed slowly underground from mafic magma.

30. What type of plate boundary is associated with the magma that formed Sample #9?
    a. ocean-ocean convergent plate boundary
    b. ocean-continent convergent plate boundary
    c. transform boundary
    d. divergent boundary

Short Answer

1. Volcanic activity creates two main types of deposits that can form igneous rocks. What are those deposits called?

2. Geologists classify pyroclastic materials according to four different particle sizes. List the four classifications and their sizes.

3. If oceanic crust is rich in ferromagnesian minerals, what type of rock would you expect to be on the ocean floor?

4. Continental crust is primarily granitic rock. What will be its mineral composition?

5. According to Bowen's reaction series, minerals do not crystallize from magma in random order, but instead crystallize based on temperature. Minerals that form mafic rocks crystallize early at high temperatures, whereas minerals that form felsic rocks crystallize later. Looking at **Figure 6.2** and using your igneous rock grid, are the minerals that form basalt early forming or late forming rock?

6. How does the way in which pumice forms make it light enough to float?

7. How does tuff form, and why does it often contain larger particles?

SEDIMENTARY ROCKS

Lesson 7

AT A GLANCE

Purpose

Learning Objectives

Materials Needed

Overview

Weathering

Clastic Sedimentary Rocks

Chemical Sedimentary Minerals and Rocks

Biochemical Sedimentary Rocks

Depositional Environments

Lab Exercises

Lab Exercise #1: *Grain Analysis*

Lab Exercise #2: *Identification of Sedimentary Rocks*

Online Activities

Quiz

Multiple Choice

Short Answer

Purpose

The activities in this lesson will lay the foundation for the understanding the processes of formation and importance of sedimentary rocks. Sedimentary rocks cover more than half of the land surface area of Earth and understanding the characteristics of sedimentary rocks is extremely important in unraveling Earth's history.

Learning Objectives

After completing this laboratory lesson, you will be able to:

- Identify textural features of sedimentary rocks.
- Explain the process of sedimentary rock formation.
- Explain how to classify sedimentary rock samples.
- Describe how sedimentary rocks reveal information about ancient environments.

Materials Needed

- ❑ 10X-power magnifying hand lens (in lab kit)
- ❑ Seven sedimentary rock specimens (in lab kit)
- ❑ Five bags of sedimentary grains labeled Bag A, Bag B, Bag C, Bag D, and Bag E (in lab kit)
- ❑ Common household lemon juice or lime juice
- ❑ White sheet of paper

Overview

Sedimentary rocks form at or near Earth's surface; therefore, sedimentary rocks provide us with a record of Earth's history. The locations of ancient beaches, rivers, deserts, glaciers, and oceans can be determined by analyzing sedimentary rocks. In addition, the vast majority of all fossils are found in sedimentary rocks, and, as will be discussed in Lesson 9, much of our understanding of the history of life is based upon the diversity and changes in the fossil record.

Economically, sedimentary rocks contain many of the natural resources needed by modern society, not the least of which is energy resources. The primary source of energy in the world today, oil and natural gas, is stored within sedimentary rock layers until retrieved by an oil or gas well. The second most important source of energy, coal, is itself considered to be a sedimentary rock. Finally, our present supplies of uranium, the fuel used in most commercial nuclear power plants, are often extracted from deposits found in sedimentary rocks. In addition to energy, sedimentary rocks provide society with a wide variety of valuable and necessary metallic commodities such as the ores of iron bauxite for aluminum. Without sedimentary rocks, modern society would have a very different look than it does today.

Sedimentary rocks form from components that once belonged to other rocks. There are three major categories of sedimentary rocks: (1) clastic, (2) chemical, and (3) biochemical. They differ in the character of their components and the ways in which they form.

Weathering

As seen in the rock cycle illustration in Lesson 6 (**Figure 6.1**), there is an intermediate step between earlier rocks and the processes that form sedimentary rocks: weathering. Weathering is the process by which rocks are broken down into smaller particles or into their chemical components.

Mechanical weathering is the type of weathering that occurs when water, wind or ice break rock down into smaller particles, called clasts. The clasts are then carried away, most often by water, in the process of erosion or through downslope movement, such as landslides, due to gravity.

Chemical weathering occurs when the soluble components of rock dissolve in water. The water then carries away these soluble components while they are in solution. The dissolution of some chemical components in rock may leave behind some insoluble components. One of the most important insoluble products of chemical weathering is left behind when orthoclase feldspar is exposed to water; the potassium in the orthoclase easily enters into solution leaving behind the less soluble silicates that were in feldspar's framework structure. These silicates then assume a new sheet structure and become part of an important group called clay minerals. *Clay minerals should not be confused with the clay-sized particles discussed below.* Clay minerals play an important role in the character of some metamorphic rocks, to be discussed in Lesson 8.

Clastic Sedimentary Rocks

Clastic sedimentary rocks are those composed of rock and mineral particles that have been cemented together. The rock particles are the result of weathering and are deposited as sediment, a term that includes both the clasts as well as loose or fragmental debris, such as leaf litter and other plant remains, and the shells of some marine organisms. All sediment has a source, or place of origin, where it was formed by physical or chemical weathering of the parent rock or by the life cycles of plants and animals. After sediment is created at the source, it is generally transported by water, ice, or wind to another location.

Changes in the energy of the water or wind carrying sediment will drop particles of various sizes along the way; for instance, a rushing stream is capable of carrying larger particles than a slow-moving stream. When a stream slows, some larger particles in the transported sediment will drop out to form a sedimentary deposit. By the time the stream reaches the ocean, it will have lost enough energy to deposit most of the sediment. A similar process of decreasing energy deposits sediments at shorelines; glaciers also pick up sediments and deposit them, although not in any particular order.

The deposition process often separates clasts according to their densities and grain sizes—a process known as sorting. A well-sorted sediment is composed of clasts that are similar sizes and/or densities whereas a poorly sorted sediment is composed of grains that vary widely in size and/or density (**Figure 7.1**). As will be discussed in Lesson 11, clasts tend to be well sorted when they've been deposited by wind, or by a body of water, such as a stream or river, in a lake or along an ocean shoreline. As discussed in Lesson 13, rocks deposited by glaciers are poorly sorted.

Sediment particle size, also referred to as grain size, is expressed in terms of it diameter:

Gravel (very coarse-grained): grains larger than 2 mm in diameter (granules, pebbles, cobbles, boulders)

Sand (coarse-grained): grains from 1/16 mm to 2 mm in diameter

Silt (fine-grained): grains from 1/256 mm to 1/16 mm in diameter

Clay (fine-grained): grains less than 1/256 mm in diameter

Most silt particles can only be seen with a hand lens. Clay-sized particles are so tiny they can only be seen with a microscope.

Grain Size	Grain Shape	Grain Arrangement
Gravel Grains visible; larger than sand	**Angular**	**Poorly Sorted**
Sand Coarse-grained: grains from 1/16 mm-2 mm in diameter	**Rounded**	**Moderately Sorted**
Silt Grains barely visible; feels gritty	**Well Rounded**	**Well Sorted**
Clay Grains not visible; feels smooth; dull luster on freshly broken surface		

Figure 7.1 Textural Features of Sedimentary Rocks. Illustration by Marie Hulett using photos as follows: (gravel) Shutterstock 62432644, credit Imageman; (sand) Shutterstock 60560716, credit jeka84; (silt) Shutterstock 41423524, credit jeka84; (clay) Shutterstock 57422938, credit dutourdumonde; (angular) Shutterstock 16544482, credit Eugene F; (rounded) Shutterstock 59134291, credit Wutthichai; (well rounded) Shutterstock 5455889, credit Olaf Speier; (poorly sorted) Shutterstock 57774088, credit javi_indy; (moderately sorted) Shutterstock 58350871, credit Karin Wabbro; (well sorted) Shutterstock 59584597, credit Mirec.

Grains are also classified by their shape. Particles are generally angular when first created by weathering, and are subject to abrasion while they are being transported. Those particles that have traveled a long distance or for a long time become well rounded, whereas particles that have not been transported very far or for very long will remain more angular. Consequently, the shape of the particle gives geologists an indication of how far or how long that particle has traveled from its source.

Once deposited, sediments may again be picked up by wind, water, or glacial ice to be further transported, or they may be buried by layers of additional deposits of sediment. If they are buried deeply enough, they undergo lithification, the process by which the clasts in sediments form clastic sedimentary rock. Lithification is a two-step process:

- **Compaction:** As sediments are buried, the weight of overlying layers creates a compressive force that pushes the clasts closer together, reducing the pore space between them.
- **Cementation:** If clasts are fine or very fine, the pressure of compaction may be enough for them to cling together to form rock. If the clasts are sand-sized or larger, they may be cemented together when dissolved minerals, typically calcite, silica, or iron oxide, crystallize and glue the clasts together.

The resulting rock body referred to as a rock bed.

Clastic sedimentary rocks are categorized by the size of the particles composing them, their shape (whether they are rounded or angular), and, to a lesser extent, their mineral composition. For instance, sandstone gains its name from the sand-sized particles that compose it. Quartz sandstone comes in a variety of colors often given to it by the minerals acting as cement; for instance, hematite (iron oxide) cement gives quartz sandstone a reddish-brown color and a cement containing manganese gives a purplish hue. Calcite cement generally results in a lighter color (and may make it fizz in an acid test!)

85 percent of all sandstone is composed of sand-sized grains of quartz. When sandstone contains less than 85 percent quartz, that fact is reflected in its name: arkose is sandstone that contains an appreciable proportion of feldspar clasts (recognizable through a hand lens by their cleavage), and graywacke is sandstone that contains a large proportion of clay minerals.

Another abundant clastic sedimentary rock is shale, which is composed of clay-sized particles of quartz and clay minerals. The clasts in shale are far too small to see, but the sheet structure of the clay minerals composing it cause it to break along parallel planes forming sheets of rock, a property called fissility. Shale is the only fissile sedimentary rock.

Figure 7.2 summarizes the most common clastic sedimentary rocks.

Rock	Clast size	Composition & Texture
Conglomerate	Gravel-sized (> 2 mm)	Well-rounded gravel-sized clasts within a matrix of cemented sand- and silt-sized particles.
Breccia	Gravel-sized (> 2 mm)	Angular, gravel-sized clasts within a matrix of cemented sand- and silt-sized particles.
Quartz Sandstone	Sand-sized (1/16 mm–2 mm)	Sand-sized clasts of quartz cemented by calcite, silica, or iron oxide.
Arkose	Sand-sized (1/16 mm–2 mm)	Sand-sized clasts of quartz and feldspar cemented by calcite, silica, or iron oxide.

Rock	Clast size	Composition & Texture
 **Graywacke**	Sand-sized (1/16 mm–2 mm)	Sand-sized clasts of quartz mixed with clay minerals and cemented by calcite, silica, or iron oxide.
 **Siltstone**	Silt-sized (1/256 mm–1/16 mm)	Silt-sized clasts of quartz and clay minerals.
 **Mudstone**	Clay-sized (< 1/256 mm)	Mixture of clay-sized particles of quartz and clay minerals compacted in such a way that they are not fissile.
 **Shale**	Clay-sized (< 1/256 mm)	Mixture of clay-sized particles of quartz and clay minerals compacted in such a way that they are fissile.

Figure 7.2 Common Clastic Sedimentary Rocks. Photos: Conglomerate, ©Science VU/Visuals Unlimited, Inc.; Breccia, Shutterstock 56061205, credit Tyler Boyes; Quartz sandstone, Shutterstock 56061262, credit Tyler Boyes; Arkose, courtesy of Denise J. Mayes; Graywacke, courtesy of David Burgess; Siltstone, Shutterstock 56068867, credit Tyler Boyes; Mudstone, Shutterstock 1312303, credit Joe Goodson; Shale, released into the public domain via Wikipedia by the photographer.

Chemical Sedimentary Minerals and Rocks

Dissolved minerals deposited in water through chemical weathering may once again come out of solution to be deposited; the result is a chemical sedimentary mineral or, if there is more than one mineral in the new deposit, a chemical sedimentary rock. In some cases, the minerals crystallize as the water in which they have been dissolved evaporates. These deposits are called evaporites; rock salt (which is mainly halite), gyprock (which is mainly gypsum), and borax are examples of evaporites.

In some cases, the dissolved minerals enter into a reaction with other substances within the water to form an insoluble substance that then drops out of the solution; this process is called precipitation and the insoluble substance is called a precipitate. The mineral calcite is one such precipitate; calcite's chemical name is calcium carbonate ($CaCO_3$). The rock formed from a mix of primarily calcite along with other minerals is called limestone. Limestone is a very common rock; it is light gray to brown in color and can be fine-grained to coarse-grained. Travertine is a beautiful form of limestone that precipitates out of water near hot springs and inside caverns. Impurities give travertine a banded appearance, and it is frequently sprinkled with tiny pores formed as the calcite hardened around organisms that lived in the hot springs.

If the calcite precipitates in agitated water, it might form layers around a grain of sand or a shell fragment to form a tiny calcite sphere called an oolite. The rock composed of oolites is called oolitic limestone. Most limestone, however, has a biochemical origin; those varieties are discussed below.

Dolomite is a common chemical sedimentary mineral that is related to limestone in that it is similar in chemical composition. Most dolomite forms as magnesium-rich groundwater reacts with previously deposited $CaCO_3$ (calcite) to form $CaMg(CO_3)_2$. When mixed with some other minerals, dolomite forms dolostone. Dolomite has many of the same properties as calcite, except it fizzes only weakly when an acid test is conducted on it, and then only if the acid is warm and the dolomite is powdered.

Silica is another dissolved substance that often precipitates out of a solution. Like calcite, silica often acts as a cement for clastic sedimentary rocks. However, also like calcite, silica can form solid masses, in which case it's called chert. Some chert is composed of microcrystalline silica, but many varieties are amorphous, meaning that there are no visible particles or crystals, even with a hand lens. Chert varieties include flint, jasper, and agate.

Figure 7.3 summarizes the properties of the most common chemical sedimentary rocks.

Rock	Texture	Composition & Properties
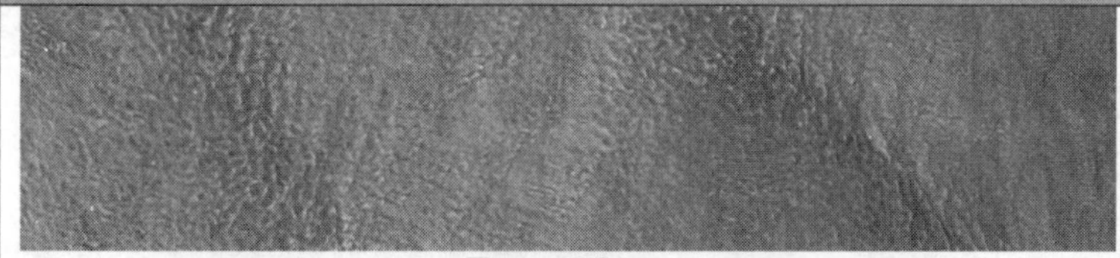 \n**Travertine**	Fine to coarse texture, crystalline	Calcite plus impurities that often give it a banded appearance. Fizzes in an acid test.
\n**Limestone**	Fine to coarse texture	Predominantly calcite as well as other minerals appearing with it. Limestone is a common rock and comes in many colors and a variety of texture. Color varies from light gray to brown. Fizzes in an acid test.
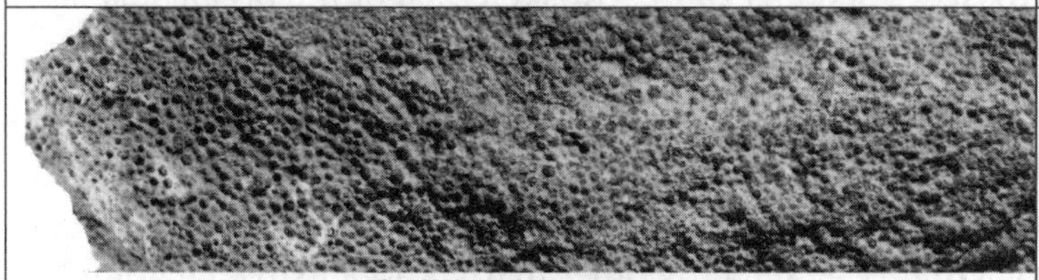 \n**Oolitic Limestone**	Coarse, spherical texture	Calcite oolites (spheres) formed around grains of sand or shell fragments. Fizzes in an acid test.
 \n**Dolostone**	Coarse texture	Composed of dolomite and impurities; fizzes weakly when powdered.
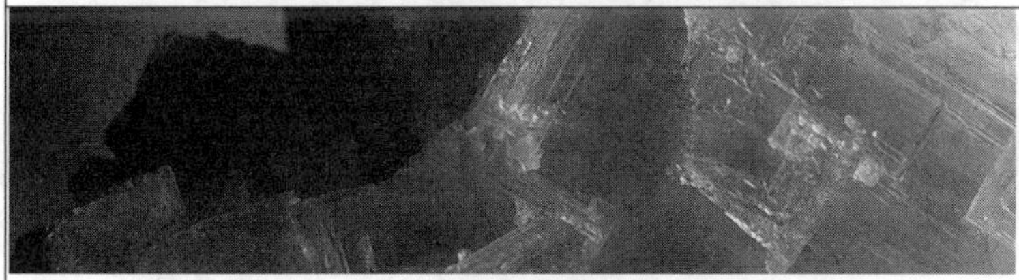 \n**Rock Salt**	Coarse texture	Halite mixed with impurities; has cleavage and salty taste of halite.
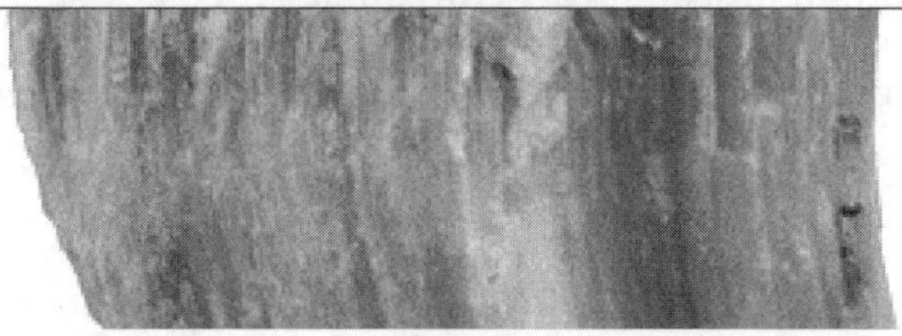 \n**Gyprock**	Fine to coarse texture	Gypsum mixed with impurities; has the cleavage and hardness of gypsum.
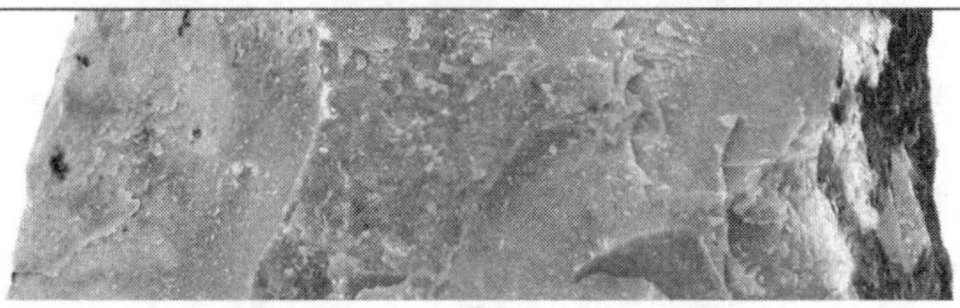 \n**Chert**	Very fine texture	Silica mixed with impurities that give it various colors; jasper, opal, chalcedony, agate, and flint are forms of chert; amorphous; has the conchoidal fracture and hardness of quartz.

Figure 7.3 Common Chemical Sedimentary Rocks. Photos: Travertine, Shutterstock 41730853, credit Oxi; Limestone, courtesy of Susan Wilcox; Oolitic limestone, courtesy of USGS; Dolostone, Shutterstock 56068813, credit Tyler Boyes; Rock salt, Shutterstock 38008114, credit Fotogiunta; Gyprock, Shutterstock 45129724, credit Terry Davis; Chert, Shutterstock 56068792, credit Tyler Boyes.

Biochemical Sedimentary Rocks

Biochemical sedimentary rocks form through much the same process as clastic sedimentary rocks; the difference is that the sediments forming them have an organic origin. Biochemical limestone, for instance, is composed of the remains of marine organisms that extract calcium carbonate from seawater in order to form their shells. When they die, the shells and other remains sink to the sea bottom where they accumulate into layers that subsequently lithify.

Most limestone deposits are biochemical in origin, and are classified based on texture and origin of the components. Chalk is a soft limestone with a very fine texture because it is composed of the remains of the fossils microscopic marine organisms. It is usually white or light gray in color, has particles of a uniformly fine size, and is not well cemented so particles easily rub off. You have probably used chalk to write on a classroom blackboard. Coquina is composed of larger shells and pieces of broken shells poorly cemented together. Fossiliferous limestone contains an abundance of fossils, generally of the exoskeletons of the organisms that secreted the calcium carbonate forming the limestone; the fossils are held together in a fine-grained matrix. Sometimes the fossils are in the form of molds that appear as tiny holes in the limestone.

Whereas limestone is composed of marine organisms that secrete calcium carbonate, biochemical chert is composed of silica that is secreted by other marine organisms. Diatoms and some other marine microorganisms form their exoskeletons from silica; layers of their remains accumulate and compact to form biochemical chert.

An important biochemical rock is coal, which forms from plant material deposited in certain swamp environments where the water is too acidic for normal decomposition to occur. Consequently, the plant materials are preserved and buried. Over time, the pressure from overlying layers compacts the deposit to form peat, a compressed form of plant materials often used for fuel. Further compaction removes some of the noncarbon substances to form lignite, a soft brown form of coal that is considered to be a sedimentary rock. Continuing compression over time as a result of even deeper burial further compacts the lignite and forces out most of the noncarbon substances to form bituminous coal, a biochemical sedimentary rock that is mostly carbon.

Figure 7.4 summarizes some of the most common biochemical sedimentary rocks.

Rock	Texture	Composition & Properties
 **Chalk**	Fine texture	Microscopic fossils and cemented skeletal remains composed of calcium carbonate; fizzes freely in an acid test.
 **Fossiliferous Limestone**	Coarse texture	Abundant invertebrate fossils large enough to be visible with a hand lens in a fine-grained calcite matrix; fizzes freely in an acid test.
 **Coquina**	Very coarse grain	Shells and shell fragments poorly cemented together by calcite; fizzes freely in an acid test.

Figure 7.4 Common Biochemical Sedimentary Rocks. Photos: Chalk, Shutterstock 56061232, credit Tyler Boyes; Fossiliferous limestone, Shutterstock 56061202, credit Tyler Boyes; Coquina, © Albert Copley/Visuals Unlimited, Inc.

Rock	Texture	Composition & Properties
 **Biochemical Chert**	Very fine grain or amorphous	Silica remains of microscopic marine organisms; has the conchoidal fracture and hardness of quartz.
 **Lignite**	Fine to coarse texture	A soft rock composed of brown to black carbonized plant remains that crumble easily and are identifiable under a hand lens.
 **Bituminous Coal**	Fine to coarse texture	Black, harder than lignite but still crumbly, with carbonized plant remains identifiable under a hand lens; conchoidal fracture.

Figure 7.4 Common Biochemical Sedimentary Rocks. (cont.) Photos: Biochemical chert, Shutterstock 40650169, credit Yury Kosourov; Lignite, courtesy of USGS; Bituminous coal, Shutterstock 56068891, credit Tyler Boyes.

Depositional Environments

Each type of sedimentary rock is formed from sediment deposited in a specific type of environment. The sediments that form clastic sedimentary rocks are most often carried by moving water and wind, then deposited when the water or wind loses energy. As discussed earlier in this lesson, streams deposit sediments as they lose energy. A rushing stream first drops the gravel-sized particles, but continues to carry the smaller particles. As it loses energy, it next drops the sand-sized particles, followed by the silt-sized particles and finally the clay-sized particles. Ocean waves and shoreline currents deposit sediments in the same order, based on their energy. Therefore, a deposit of well-sorted gravels speaks of a high-energy depositional environment; perhaps a rushing stream or an energetic shoreline. Well-sorted silt- or clay-sized particles speak of calm waters with low enough energy for these particles to settle out; perhaps a calm lake, marsh, or lagoon. Poorly sorted deposits, as noted earlier, are typical of deposits left behind by glaciers. Or they may have been deposited by a landslide or, perhaps, a turbidity current, an ocean current that transports large amounts of sediment down an underwater slope. In each of the latter cases, the sediment has been quickly dumped.

Clastic sedimentary rocks may also preserve features created by the environment in which they formed. Waves in shallow water can create ripples that are subsequently preserved in a sedimentary layer; winds and water can leave marks called cross bedding in the deposit that extends parallel to the direction of their flow; mud composed of silt- or clay-sized particles can dry and crack only to be covered by another layer that then preserves the mud cracks. Learning to notice, recognize, and interpret such structures is one of the skills geologists apply in reading rocks to determine the history of an area.

Chemical and biochemical sedimentary rocks also leave clues to the past. Evaporites form when a body of water dries up, leaving behind previously dissolved minerals; thus a rock salt or gyprock deposit speaks of the former presence of ancient sea or lake. Biochemical limestone forms at the bottom of warm, shallow seas or along shorelines, whereas biochemical chert can form from organisms that prefer colder waters.

Here is a partial summary of what sedimentary rocks can reveal about an area's history:

- Limestone most often indicates that the area was once a warm, shallow ocean.
- Chert can indicate the former presence of an ocean or a deep lake.
- Shale indicates that the area was once covered by waters calm enough for fine sediments to deposit; mudflats or deep ocean are two environments in which fine particles may settle.
- Coal or organic-rich shale indicate the area was once a swamp.
- Sandstone tells a geologist that the area was probably the site of a river, a beach, or desert, since that is where large amounts of sand accumulate. Arkose would be typical of a desert environment because it forms in dry environments.
- Cross bedding in different directions indicates a wind deposit of sand, typical of a desert environment or sand dune at a beach.
- Cross bedding in a consistent direction indicates a stream deposit; comparing cross beds across a region can help geologist to reconstruct an ancient river or delta system.
- Symmetrical ripple marks indicate a shallow coastal area, whereas asymmetrical ripple marks point to deposition in a flowing stream.

- Mud cracks tell a geologist that the area was probably a warm, dry region next to an ocean or an inland lake that periodically flooded and then dried.
- Graded beds containing layers of smaller and smaller well-sorted clasts can indicate a former lake bed, or perhaps an area of the ocean floor that was later uplifted.
- Conglomerate indicates a high-energy environment, like a rushing stream or beach with pounding waves. The energy had to be high enough for gravels to be transported to the area.
- Evaporites can indicate the former presence of a shallow ocean or an isolated basin, like a desert lake, that subsequently dried up.

Fossils add to the clues provided by sedimentary rocks. As will be discussed in Lesson 9, fossils record Earth's history and the types of environments that existed in a region over time.

Lab Exercises

Lab Exercise #1: *Grain Analysis*

Clastic sedimentary rocks are formed from clasts, also called grains, that are deposited in a variety of environments. The environment determines not only the size of the clasts deposited but also whether they are well sorted or poorly sorted.

In this laboratory exercise, you'll lay a foundation for the identification of clastic sedimentary rocks by analyzing samples of sediment (Bags A through E in your lab kit) for grain size and sorting. In doing so, you'll learn how to better determine grain size in clastic sedimentary rocks, a key clue to their identity.

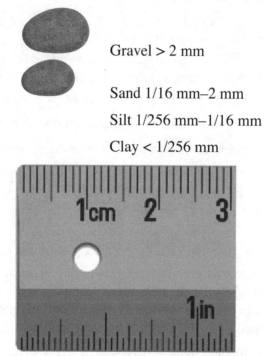

Gravel > 2 mm

Sand 1/16 mm–2 mm

Silt 1/256 mm–1/16 mm

Clay < 1/256 mm

Figure 7.5 Relative Clast Size (Enlarged). Photos: Rock image, Shutterstock 76689523, credit schankz; Ruler showing clast size, Shutterstock 72147730, credit Alhovik.

Figure 7.5 above is an enlarged illustration that will give you an idea of the relative size of the smallest possible gravel clast compared to a typical sand grain. Note that even though **Figure 7.5** is enlarged, silt- and clay-sized particles are still too small to be seen.

The ruler in your lab kit includes markings for millimeters. Use the ruler to determine the clast size of visible grains, using your hand lens as necessary. When you sort your clast samples, you will need to use the hand lens to measure silt particles. It's possible to isolate silt-sized particles by spreading a sample on the white paper, then pouring the sample back into the bag. The "dust" left on the white paper will be mostly silt-sized.

Clay-sized particles can only be seen through a microscope. One way to identify silt- and clay-sized particles is by feel. Silt-sized clasts will feel gritty whereas clay-sized particles will feel smooth and silky, descriptions that will also apply to the sedimentary rocks you will identify in **Lab Exercise #2**.

As you study each bag of sedimentary clasts, record your observations in the appropriate cells of the chart **Figure 7.6** or on a separate piece of paper. Make sure to save your results. You will use the data to answer the questions at the end of this lesson.

Instructions

Step 1: Retrieve Bag A, Bag B, Bag C, Bag D, and Bag E from your lab kit. Each bag contains a different sample of sedimentary clasts.

Step 2: Place a white sheet of paper on a flat surface.

Step 3: Pour a sample of the contents of Bag A on the white sheet of paper.

Step 4: Analyze the sediment sample for clast size. Separate out clasts that are representative of the largest and smallest sizes, then measure them, if possible, using the hand lens as necessary. Enter the range into Column 1 of **Figure 7.6**.

Step 5. Analyze the sediment sample for clast shape. Examine at least five clasts, using your hand lens if necessary, and referring to **Figure 7.1 Textural Features of Sedimentary Rocks** to help you. Are the clasts well-rounded, rounded, or angular? Record your observation in Column 2 of **Figure 7.6**.

Step 6: Analyze the sediment sample for sorting. Are the clasts about the same size? Or is there a wide range of sizes? Determine whether the sample is well-sorted, moderately sorted or poorly sorted, referring to **Figure 7.1** to help you. Record your observation in Column 3 of **Figure 7.6**.

Step 7: Determine whether the sample is clay, silt, sand, gravel, or some combination of these. Enter your determination in Column 4 of **Figure 7.6**.

Step 8: Repeat the process for Bag B, Bag C, Bag D, and then Bag E. Record your observations in **Figure 7.6** below.

Sample	Column 1: Clast Size	Column 2: Clast Shape	Column 3: Sorting	Column 4: Identification
Bag A				
Bag B				
Bag C				
Bag D				
Bag E				

Figure 7.6 Grain Analysis Chart.

Lab Exercise #2: *Identification of Sedimentary Rocks*

In this laboratory exercise, you will identify some common sedimentary rocks based on their texture and composition. Feel free to download the Sedimentary Rocks Photo Guide from the online supplement to help you make your identification.

Instructions

Step 1: Retrieve the bag labeled Lab #7 Sedimentary Rock Samples from your lab kit, and place the specimens (numbered 11 through 17) on a white sheet of paper.

Step 2: Determine the identity of each rock, then write its name in the proper cell of **Figure 7.7**.

The flow chart in **Figure 7.8** will help you to narrow down the possibilities. Feel free to conduct tests you learned in Lesson 5 to help you identify mineral components. If you decide to conduct an acid test, rinse the sample in water and dry it first so that you don't get a false positive result from dust in the bag that may have settled on the sample.

Sample #	Identification
#11	
#12	
#13	
#14	
#15	
#16	
#17	

Figure 7.7 Sedimentary Rock Identification Chart.

Sedimentary Rock ID Flow Chart

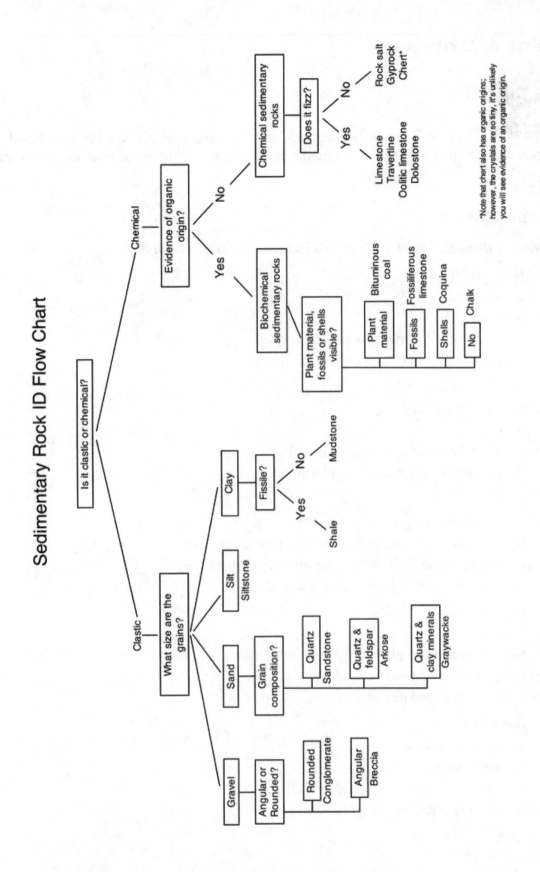

Figure 7.8 Sedimentary Rock Identification Flow Chart. Illustration by Susan Wilcox.

Online Activities

Per your instructor's directions, go to the online supplement for this lab and complete the activities assigned. Viewing the online videos will help you to complete the quiz.

Quiz

Multiple Choice

Questions 1 through 5 are based on **Lab Exercise #1:** *Grain Analysis.*

1. Bag A contains
   a. well-sorted sand.
   b. clay.
   c. a moderately sorted mix of sand and silt.
   d. a poorly sorted mix of gravel and sand.

2. Bag B contains
   a. well-sorted sand.
   b. clay.
   c. a poorly sorted mix of sand and silt.
   d. a poorly sorted mix of gravel and sand.

3. Bag C contains
   a. well-sorted clay.
   b. well-sorted gravel (with some sand from transport).
   c. poorly sorted sand (with some silt from transport).
   d. a poorly sorted mix of gravel and sand.

4. Bag D contains
   a. clay.
   b. well-sorted gravel.
   c. a poorly sorted mix of sand and silt.
   d. a poorly sorted mix of gravel and sand.

5. Bag E contains
   a. well-sorted sand.
   b. well-sorted gravel.
   c. a poorly sorted mix of sand and silt.
   d. a poorly sorted mix of gravel, sand, and silt.

6. The mineral grains in Specimen #11 are
    a. biotite mica.
    b. feldspar.
    c. quartz.
    d. hornblende.

7. Specimen #11 is
    a. chert.
    b. shale.
    c. quartz sandstone.
    d. arkose.

8. Specimen #12 has black carbonized plant remains. What is specimen #12?
    a. fossiliferous limestone
    b. bituminous coal
    c. graywacke
    d. rock salt

9. Based on the clasts in Specimen #13, the deposition environment might have been
    a. sand dunes.
    b. a warm, shallow lagoon.
    c. a lake bottom.
    d. a river channel.

10. The largest grain-size dimension for Specimen #13 is
    a. greater than 2 mm in diameter.
    b. between 1/16 mm and 2 mm in diameter.
    c. between 1/256 mm and 1/16 mm in diameter.
    d. less than 1/256 mm in diameter.

11. Specimen #13 is
    a. chert.
    b. shale.
    c. fossiliferous limestone.
    d. conglomerate.

12. Specimen #14 contains obvious
    a. plant material.
    b. mammalian fossils.
    c. fossil molds from tiny invertebrates.
    d. mollusk fossils.

13. Specimen #14 is
    a. sandstone.
    b. conglomerate.
    c. fossiliferous limestone.
    d. gyprock.

14. Specimen #15
    a. is fissile.
    b. contains visible shell fragments.
    c. fizzes during an acid test.
    d. feels gritty to the touch.

15. Specimen #15 is
    a. quartz chert.
    b. shale.
    c. limestone.
    d. conglomerate.

16. Which of the following is a diagnostic feature of Specimen #17?
    a. It fizzes when tested with acid.
    b. It is a very coarse-grained sedimentary rock.
    c. It is a fissile sedimentary rock.
    d. It is obviously the product of a high-energy depositional environment.

17. Specimen #17 is
    a. conglomerate.
    b. gyprock.
    c. quartz chert.
    d. shale.

18. Specimen #16 is composed of
    a. silica.
    b. calcium carbonate.
    c. grains of sand-sized particles.
    d. gypsum.

19. Which of the following is a diagnostic feature of Specimen #16?
    a. It effervesces when tested with acid.
    b. It is a very coarse-grained sedimentary rock.
    c. It is a fine-grained sedimentary rock.
    d. It is amorphous.

20. Specimen #16 is
    a. shale.
    b. chalk.
    c. sandstone.
    d. chert.

21. _____ gives quartz sandstone its color.
    a. The mineral content of its clasts
    b. The mineral content of its cement
    c. Fossils
    d. Plant remains

22. The most common chemical sedimentary rocks are largely composed of what type of mineral?
    a. calcite and silica
    b. gypsum and halite
    c. borax and gypsum
    d. quartz and borax

23. The process that breaks rock down into particles or its chemical components is called
    a. compaction.
    b. cementation.
    c. weathering.
    d. lithification.

24. The process by which soluble minerals are dissolved by water is called
    a. mechanical weathering.
    b. chemical weathering.
    c. precipitation.
    d. lithification.

25. The two processes by which chemical sedimentary rocks are deposited are
    a. compaction and cementation.
    b. evaporation and precipitation.
    c. erosion and transportation.
    d. weathering and lithification.

26. The fine sediments that form shale are typically deposited
    a. by the surf at a beach.
    b. as a glacier retreats.
    c. in a high-energy environment like a rushing stream.
    d. in a low-energy environment like a calm lake.

27. How can one distinguish breccia from conglomerate?
    a. Breccia contains angular clasts and conglomerate contains rounded clasts.
    b. Breccia contains rounded clasts and conglomerate contains angular clasts.
    c. Different minerals cement breccia than cement conglomerate.
    d. Breccia contains sand and conglomerate contains silt.

Short Answer

1. When rocks weather into particles, most of the rocks don't stay in place. Name three ways that weathered rock particles are carried away.

2. Geologists classify sedimentary rock particles based upon their grain size. List the four sizes of particles in order from smallest to largest.

3. Name the two processes by which sediment is converted into rock and provide a definition for each process.

4. Explain the sequence that produces the sedimentary rock, bituminous coal.

5. Give two examples of evaporite minerals, and explain how they are deposited.

6. Explain why limestone can be considered both a chemical and biochemical sedimentary rock.

METAMORPHISM AND METAMORPHIC ROCKS

Lesson 8

AT A GLANCE

Purpose

Learning Objectives

Materials Needed

Overview

> Metamorphic Conditions
>
> Metamorphic Processes and Settings
>
> Metamorphic Rock Identification

Lab Exercise

> Lab Exercise: *Identification of Metamorphic Rocks*

Online Activities

Quiz

> Multiple Choice
>
> Short Answer

Purpose

The activities in this lesson will lay the foundation for understanding the processes by which elevated levels of heat, pressure, and chemically active fluids in Earth's lower crust and upper mantle create metamorphic rocks. The history most metamorphic rocks reveal is of a journey deep beneath Earth's surface where temperatures and pressures create transformations that occur nowhere else on Earth.

Learning Objectives

After completing this laboratory lesson, you will be able to:

- Explain the origin of metamorphic rocks.
- Describe the processes by which metamorphic rocks are created.
- Characterize the different textural features of metamorphic rocks.
- Use physical properties to identify metamorphic rock samples.

Materials Needed

- ❏ Common household lemon juice or lime juice
- ❏ 10 X-magnifying hand lens (in lab kit)
- ❏ Metamorphic rock specimens (in lab kit)
- ❏ White sheet of paper

Overview

Like igneous and sedimentary rock, metamorphic rocks have a history to reveal.

Metamorphic rocks are those that have changed their character as a result of extreme conditions. Under such conditions, they become denser, and the minerals they contain rearrange their atoms so that their crystals are larger. Often, the minerals within the rock swap atoms to form new minerals altogether. The resulting metamorphic rock may bear little resemblance to the original parent rock.

Such dramatic changes can only happen under extreme conditions; most metamorphic rocks undergo these changes deep beneath Earth's surface and near convergent boundaries. Thus, when a geologist encounters a metamorphic rock body, it is usually an indication that the region once underwent tremendous stresses that buried rocks and subsequently uplifted them once again to the surface.

Metamorphism occurs because minerals do not remain stable—that is, maintain their unique chemical structures—under extreme conditions. Minerals are stable only under a range of conditions close to those in which they originally crystallized. When temperatures or pressures rise, the minerals respond with the dramatic changes described in this lesson. The changes in the minerals transform the identity and character of the rocks they compose.

How dramatic the transformation is and the type of new rock that results depend both upon the parent rock, or protolith, and the intensity of the conditions. Different scenarios produce different types of metamorphic conditions.

Metamorphic Conditions

Metamorphism takes place when a rock is subjected to increases in temperature and/or pressure or contact with a hot, chemically active fluid, usually hot water. The type of changes vary with the intensity of the conditions.

As discussed in earlier lessons, temperature and pressure increases with depth below Earth's surface. Combinations of heat and pressure that produce metamorphism are called metamorphic grades. The higher the metamorphic grade, the more the metamorphic rock will differ from its protolith.

Figure 8.1 shows the temperature and pressure combinations that produce low-grade, intermediate-grade, and high-grade metamorphic conditions.

- Low-grade metamorphic conditions are the least extreme, and produce metamorphic rocks that are only slightly changed in texture and composition from their protoliths.

- Intermediate-grade metamorphic conditions produce rocks that have a greater change in texture and composition than those produced in low-grade conditions because they are transformed under more extreme conditions: moderate temperatures with low to moderate pressures, or low temperatures at high pressures.

- High-grade metamorphic conditions produce the greatest change in the texture and composition of rocks. High-grade conditions feature either high temperatures or high pressures, and often a combination of both.

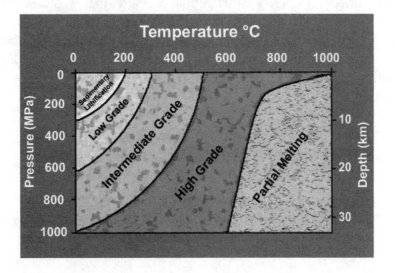

Figure 8.1 Metamorphic Grades. Metamorphic grades are based on pressure/temperature combinations. The top limit of metamorphism is at the pressure/temperature combination at which melting begins. Illustration by Marie Hulett.

There is a limit, however, to how extreme metamorphic conditions can become before the rocks begin to melt. Once melting occurs, the rock becomes magma, and any new rock that crystallizes from the magma will be, by definition, igneous.

Metamorphic Processes and Settings

Compaction and Recrystallization

When any rock or mineral is buried or carried deeply beneath Earth's surface, the increasing pressure squeezes its grains and the atoms in its atomic structure closer together, making the rock or mineral denser. As pressure and temperature continue to increase and conditions become more extreme, the minerals composing the rock begin to lose their stability. Their components begin to adjust their positions until they find a new structure that is stable under the current conditions.

The most common change is recrystallization, or the slow conversion of small crystals of one mineral into larger crystals of the same mineral. For example, microscopic muscovite crystals in the metamorphic rock slate can progressively grow, or recrystallize, as they are subjected to progressively more extreme conditions. They may become the larger muscovite crystals typical of the metamorphic rocks phyllite and schist.

If the protolith is chemically complex, metamorphism can convert preexisting minerals into new minerals by the recombination of the same chemical elements. For example, the sedimentary rock shale generally contains an abundance of clay minerals, but clay minerals vary widely in chemical composition. Therefore, clay minerals provide a large reservoir of silicon, potassium, sodium, iron, magnesium, and calcium that can be recombined into many different minerals.

Foliation

Some metamorphic settings produce differential pressure; that is, pressure greater in one direction than in another. If the grains within the rock are elongate—that is, longer in one dimension than another—differential pressure will cause them to adjust their positions so that they line up in the same direction, one that is perpendicular to the force (**Figure 8.2**). The new direction is called their preferred orientation.

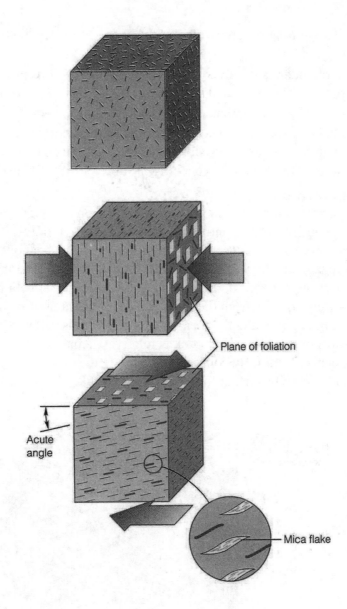

Figure 8.2 Foliation. Elongate crystals and platy minerals, like mica, rotate under differential pressure into a preferred orientation. The middle image shows the effect of compression, whereas the bottom image shows the direction of rotation under shear stress. Preferred orientation produces a property called foliation. Illustration by John J. Renton.

Minerals with elongate grains include some members of the amphibole group. Minerals with sheet structures, like mica, are called platy minerals and also adopt a preferred orientation under differential stress. In both cases, their preferred orientation results in a property called foliation, and the rocks they form are referred to as foliated rocks. Foliated rocks are an important category of metamorphic rocks and are further discussed later in this lesson. Metamorphic rocks that do not have grains lined up in a preferred orientation are called nonfoliated rocks.

Compaction, recrystallization, and foliation occur when rocks are carried or buried deeply enough underground for pressures and temperatures to become high enough for metamorphism to occur. Three settings produce such conditions:

- Subduction zones where a subducting plate slides under another plate, carrying rocks deep into Earth, where they are subjected to increasingly high temperatures and pressures. This setting progressively produces low-, intermediate-, and high-grade metamorphic conditions.

- Convergent boundaries where continent-continent collisions produce compressive forces that push rock far above and below Earth's surface. This setting also progressively produces low-, intermediate-, and high-grade metamorphic conditions.
- Some sedimentary rock settings where accumulating sedimentary layers can bury rocks deeply enough for the pressures and temperatures to reach those of metamorphic conditions. This type of setting generally produces low-grade metamorphism.

Contact Metamorphism

Contact metamorphism occurs when molten rock makes contact with preexisting rock, a common occurrence during igneous intrusions and beneath lava flows. Some of the crystalline structures in the country rock are unstable in the presence of the heat of magma, and so they metamorphose to regain stability. As the heat from the molten rock permeates the surrounding "country rock," it creates zones of high-, medium-, and low-grade metamorphism as temperatures drop with distance from the magma. The entire area of metamorphosed rock is called an aureole (**Figure 8.3**), and the rock that composes it is called hornfels, regardless of the parent rock. If the volume of the molten rock is relatively small, as it is under a lava flow, the aureole might be relatively thin. By contrast, the aureole that that forms with the intrusion of a huge mass of magma could be tens or hundreds of feet thick.

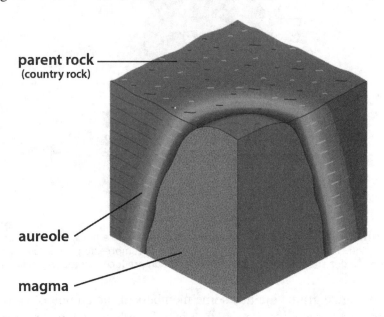

Figure 8.3 Contact Metamorphism. The country rock immediately adjacent to an intruding mass of magma is altered to form a zone of metamorphism called an aureole. The thickness of the aureole depends on the amount of heat available from the mass of the intruding magma. Illustration by John J. Renton.

Hydrothermal Processes

Hydrothermal metamorphism occurs when rock makes contact with a hot, chemically active fluid (hot water). In this case, some of the dissolved substances in the hot water may swap places with some of the components in the chemical structure of minerals in the surrounding rocks. When this happens, the swap alters the chemical composition of the minerals, which changes them into different minerals. The resulting new composition produces a metamorphic rock that differs in both composition and texture from the parent rock.

In some cases, all of the components of the rock's original minerals are replaced by components donated by the water. This process of complete replacement is called metasomatism, and it produces most of the copper, gold, zinc, and other deposits found in metamorphic rocks.

Metamorphic Rock Identification

The two major groups of metamorphic rocks are those that are foliated and those that are nonfoliated.

Foliated Metamorphic Rocks

As noted earlier, foliated metamorphic rocks form when rocks containing elongate crystals, are subjected to metamorphic conditions. The most common protolith for foliated metamorphic rock is the abundant sedimentary rock shale, although other rocks, like granite, that contain platy minerals are also likely to become foliated if subjected to metamorphic conditions. The changes that these rocks undergo forms a sequence that produces (in order of increasing grade) slate, phyllite, schist, and gneiss, based on the texture that results from being exposed to increasingly intense metamorphic conditions.

Slate

Under low-grade metamorphic conditions, shale's clay minerals recrystallize to form quartz, mica, and chlorite, an important dark-green platy mineral that is a major component of many metamorphic rocks, including slate (**Figure 8.4**). Slate appears as gray, greenish-gray, and in various shades of red. Slate is extremely fine-grained, which makes it relatively impervious to water, one reason it is often used as a roofing material. Another is slate's exceptional cleavage. Slate cleaves so cleanly in a single plane that it is the preferred material for fine pool tables, which must be absolutely smooth and flat. This type of cleavage is so associated with slate that metamorphic rocks with similar cleavage are said to have slaty cleavage. Slate's cleavage, however, is exceptionally clean compared to other metamorphic rocks with slaty cleavage.

Figure 8.4 Slate. Slate is a fine-grained foliated rock recognizable by its exceptionally clean slaty cleavage. Photo: Shutterstock 56061259, credit Tyler Boyes.

Phyllite

The continued metamorphism of slate in low-grade conditions produces phyllite (**Figure 8.5**). Like slate, phyllite contains platy minerals; however, under the more intense metamorphic conditions, the crystals in phyllite grow larger than those in slate, although they are still microscopic. Their presence is revealed by phyllite's coarser texture and more irregular surface compared to slate; phyllite is often described as having a "crinkly" appearance and a satiny sheen much like frosted eye shadow (which contains phyllite). In addition, while phyllite cleaves in one plane, it does not cleave as smoothly as slate.

Figure 8.5 Phyllite. The larger crystals in phyllite give this rock a crinkly surface and a sheen similar to that of frosted eye shadow. Photo: © Marli Miller/Visuals Unlimited, Inc.

Schist

When pressure and heat reach intermediate-grade conditions, the same platy minerals contained in slate and phyllite transform into schist (**Figure 8.6**). Under these conditions, the chlorite breaks down, and the atoms rearrange themselves to form a number of other minerals, including mica and the nonplaty minerals feldspar and quartz. The crystals in schist are much larger than the crystals in slate or phyllite, large enough to be seen with the naked eye and easily seen with a hand lens.

Figure 8.6 Schist. Schist is an intermediate-grade metamorphic rock that has foliated crystals large enough to see without magnification, giving it a shiny appearance. Photo: Released into the public domain by Daniel Engelhardt via Wikipedia.

Schist's larger crystals and abundance of nonfoliated minerals results in an uneven distribution of minerals, with the mafic and felsic components beginning to separate from each other. The separation often gives schist light (felsic) and dark (mafic) streaks, and light reflecting off its larger crystals gives schist a relatively shiny appearance. The uneven composition keeps schist from being able to cleave in a single plane, as slate and phyllite do; rather, the minerals in schist tend to break off into flakes. In fact, the term *schist* comes from the Greek word meaning "to split."

Gneiss

Gneiss forms when the blend of mica, feldspar, and quartz found in schist (and many granites) are subjected to high-grade metamorphic conditions (**Figure 8.7**). Under these conditions, the felsic and mafic crystals in the rock further segregate to form light and dark parallel bands that are the defining feature of gneiss. In fact, this segregation is so typical of gneiss that it is called gneissic banding; it makes gneiss one of the easiest rocks to identify.

Figure 8.7 Gneiss. The high-grade foliated metamorphic rock gneiss is easily recognizable by its characteristic banding, formed when mafic and felsic minerals separate. Photo: Shutterstock 56061208, credit Tyler Boyes.

Gneiss is the most common of all metamorphic rocks because it can form from a wide and diverse variety of protoliths. Common mineral constituents include amphibole, quartz, orthoclase feldspar, and plagioclase feldspar.

Identifying Foliation

Even though foliation is a key property in metamorphic rock identification, it's not always easy to describe or to recognize. There are several ways to identify a rock sample as foliated, depending on the size of the grain.

- If the grain is too fine to see with the naked eye or with a magnifying glass, look for slaty cleavage. If the rock has slaty cleavage, it is foliated.
- Fine-grained foliated rocks can have a wavy or crinkled appearance that is not seen in igneous, sedimentary rocks, or nonfoliated metamorphic rocks.
- If the grains are large enough to see, a preferred orientation of the grains may be visible, particularly with a hand lens. Only foliated metamorphic rocks have preferred orientations.
- If there is banding, the rock is, in most cases, foliated. In some cases, nonfoliated rocks, like marble, can show signs of light banding. If there is doubt as to whether the rock is foliated or a lightly banded nonfoliated rock, other tests like hardness or acid will help you to make the determination.

Nonfoliated Metamorphic Rocks

Nonfoliated metamorphic rocks do not have directional properties because they don't contain elongate crystals. The most common nonfoliated metamorphic rocks are quartzite, marble, hornfels, serpentinite, and anthracite coal.

When sandstone is subjected to metamorphic conditions, it forms quartzite (**Figure 8.8**). Quartzite is a dense, hard, generally light-colored, metamorphic rock composed of recrystallized quartz grains that have been solidly fused together; when quartzite fractures, it breaks through rather than around the fused grains. Quartzite is denser than quartz, and, like quartz, has a hardness of 7; it can readily scratch a glass plate. A scratch test is one way to tell quartzite from marble, which often has a similar surface appearance.

Figure 8.8 Quartzite. Quartzite is a nonfoliated rock that forms when sandstone is subjected to metamorphic conditions. Quartzite fractures rather than cleaves because its crystals are interlocked. Photo: Shutterstock 56061265, credit Tyler Boyes.

Marble forms when rocks containing calcite, such as travertine or limestone, are subjected to metamorphic conditions (**Figure 8.9**). Marble is a medium- to coarse-grained, variably colored, metamorphic rock composed of tightly interlocking uniform grains of recrystallized calcite, which give it an even texture. Because it's composed of calcite, marble effervesces during an acid test, although not nearly as energetically as calcite. Like calcite, marble also has a hardness of 3 and, therefore, cannot scratch a glass plate.

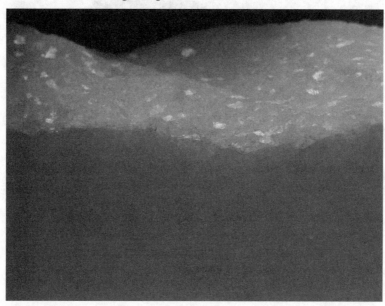

Figure 8.9 Marble. Marble forms when calcite-based rocks, like limestone and travertine, are subjected to metamorphic conditions. Like quartzite, marble has interlocking crystals. Photo: Released into the public domain by the photographer via Wikipedia.

Serpentinite is a generally fine-grained, mottled dark and light green, metamorphic rock composed chiefly of the mineral serpentine (**Figure 8.10**). Its parent rock is the ultramafic igneous rock peridotite, and it forms through hydrothermal processes. Serpentinite sometimes appears moderately foliated; however, it does not have the slaty cleavage of fine-grained foliated rocks.

Figure 8.10 Serpentinite. Serpentinite is a metamorphic rock formed through hydrothermal processes. In this sample, the serpentinite is the production of a reaction between seawater and peridotite that has been brought to the surface from deep within the mantle by tectonic processes. The thin fractures in the serpentinite are filled with precipitated calcium carbonate. Photo: Courtesy of G. Fruh-Green/NOAA.

Anthracite coal is a hard, fine-grained, shiny black metamorphic rock, also called hard coal (**Figure 8.11**). Like its protolith, bituminous coal, anthracite is composed of carbonized plant remains, but it is denser than bituminous coal and cannot be easily broken apart. Anthracite coal has a homogeneous texture and often breaks along glossy, conchoidal fractures.

Figure 8.11 Anthracite. Anthracite is the metamorphic version of coal, and is denser, harder and shinier than the sedimentary rock, bituminous coal. Photo: Courtesy of Marie Hulett.

Hornfels is a dense, fine-grained, dark-colored, metamorphic rock (**Figure 8.12**) that results from contact metamorphism. Its mineralogical composition varies depending upon that of its protolith. Hornfels is recognizable by its fine grain, dark color, and baked appearance.

Figure 8.12 Hornfels. Hornfels is a hard, dense rock that forms through contact metamorphism and is recognizable by a baked appearance. Photo: Courtesy of NASA.

Lab Exercise

Lab Exercise: *Identification of Metamorphic Rocks*

In this laboratory exercise, you will identify the most common metamorphic rocks based on their texture and composition. Feel free to download the Metamorphic Rocks Photo Guide from the online supplement to help you make your identification.

Instructions

Step 1: Retrieve the bag labeled Lab #8 Metamorphic Rock Samples from your lab kit, and place the specimens (numbered 18 through 25) on a white sheet of paper.

Step 2: Determine the identity of each rock, then write its name in the proper cell of **Figure 8.13**.

The flow chart in **Figure 8.14** will help you to narrow down the possibilities. The first determination to make is whether a sample is foliated or nonfoliated. Remember that some nonfoliated rocks occasionally show signs of banding. If you suspect a rock is not really foliated, follow up by looking for other properties, by identifying component minerals, and by conducting other tests, as necessary. If you decide to conduct an acid test, rinse the sample in water and dry it first so that you don't get a false positive result from dust in the bag that may have settled on the sample.

Sample #	Identification
#18	
#19	
#20	
#21	
#22	
#23	
#24	
#25	

Figure 8.13 Metamorphic Rock Identification Chart.

Metamorphic Rock ID Flow Chart

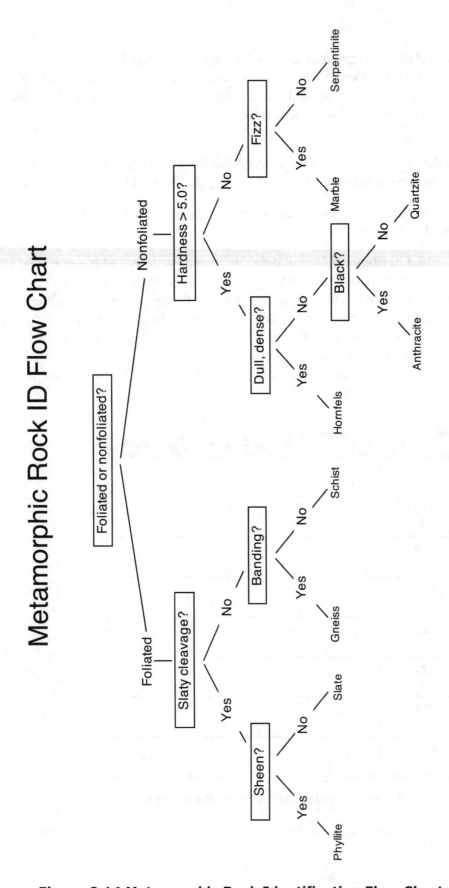

Figure 8.14 Metamorphic Rock Identification Flow Chart. Illustration by Susan Wilcox.

Online Activities

Per your instructor's directions, go to the online lesson for this lab and complete the activities assigned. Viewing the online videos will help you to complete the quiz.

Quiz

Multiple Choice

Questions 1 through 23 are based on the **Lab Exercise:** *Identification of Metamorphic Rocks. Refer to your results from the lab exercise and the Overview to assist you in answering the questions.*

1. Specimen #18 formed under which metamorphic conditions?
   a. contact metamorphism
   b. low-grade metamorphic conditions
   c. intermediate-grade metamorphic conditions
   d. high-grade metamorphic conditions

2. What property is diagnostic for specimen #18?
   a. banding
   b. slaty cleavage
   c. crinkly appearance
   d. effervescence during acid test

3. Specimen #18 is
   a. gneiss.
   b. schist.
   c. phyllite.
   d. slate.

4. Specimen #19 formed under which metamorphic conditions?
   a. contact metamorphism
   b. low-grade metamorphic conditions
   c. intermediate-grade metamorphic conditions
   d. high-grade metamorphic conditions

5. What property is diagnostic for specimen #19?
   a. slaty cleavage
   b. visible crystals
   c. satiny sheen
   d. fizzes during acid test

6. Specimen #19 is
    a. gneiss.
    b. schist.
    c. phyllite.
    d. slate.

7. Specimen #20 formed under which metamorphic conditions?
    a. contact metamorphism
    b. low-grade metamorphic conditions
    c. intermediate-grade metamorphic conditions
    d. high-grade metamorphic conditions

8. What property is diagnostic for specimen #20?
    a. exceptionally clean slaty cleavage
    b. visible crystals
    c. satiny sheen
    d. fizzes during acid test

9. Specimen #20 is
    a. gneiss.
    b. schist.
    c. phyllite.
    d. slate.

10. Specimen #21 is
    a. a low-grade metamorphic rock.
    b. a low- to intermediate-grade metamorphic rock.
    c. an intermediate- to high-grade metamorphic rock.
    d. a high-grade metamorphic rock.

11. What property is diagnostic for specimen #21?
    a. exceptionally clean slaty cleavage
    b. visible crystals
    c. satiny sheen
    d. fizzes during acid test

12. Specimen #21 is
    a. gneiss.
    b. schist.
    c. phyllite.
    d. slate.

13. The protolith for specimen #22 might have been
    a. rock salt.
    b. limestone.
    c. sandstone.
    d. shale.

14. What property is diagnostic for specimen #22?
    a. exceptionally clean slaty cleavage
    b. visible crystals
    c. satiny sheen
    d. fizzes during acid test

15. Specimen #22 is
    a. marble.
    b. quartzite.
    c. hornfels.
    d. anthracite coal.

16. The precursor for specimen #23 might have been
    a. rock salt.
    b. limestone.
    c. sandstone.
    d. shale.

17. Specimen #23
    a. is harder than glass.
    b. effervesces when treated with acid.
    c. is foliated.
    d. has a dull luster.

18. Specimen #23 is
    a. marble.
    b. quartzite.
    c. hornfels.
    d. anthracite coal.

19. Specimen #24 formed under which metamorphic conditions?
    a. contact metamorphism
    b. low-grade metamorphic conditions
    c. intermediate-grade metamorphic conditions
    d. high-grade metamorphic conditions

20. Specimen #24

    a. has a baked appearance.
    b. effervesces when treated with acid.
    c. is foliated.
    d. is banded.

21. Specimen #24 is

    a. marble.
    b. quartzite.
    c. hornfels.
    d. anthracite coal.

22. The precursor for specimen #25 might have been

    a. shale.
    b. limestone.
    c. basalt.
    d. bituminous coal.

23. Specimen #25 is

    a. marble.
    b. quartzite.
    c. hornfels.
    d. anthracite coal.

24. Under differential pressure, elongate mineral grains

    a. line up in a preferred orientation.
    b. begin to melt.
    c. fuse together.
    d. begin to dissolve.

25. Which is **NOT** an indication that a rock is foliated?

    a. slaty cleavage
    b. fused grains
    c. banding
    d. a wavy or crinkled appearance

26. Quartzite is _____ than quartz.

    a. harder
    b. clearer
    c. more foliated
    d. denser

27. Which is **NOT** a metamorphic setting?
    a. the area around a magma intrusion
    b. a mountain valley
    c. a subduction zone
    d. a continent-continent collision

Short Answer

1. What is metamorphism?

2. How is foliation different than sedimentary layering?

3. What properties of slate make it good roofing material?

4. What is a metamorphic grade? What metamorphic grades are associated with the four main categories of foliated rocks?

5. What history does the presence of most metamorphic rocks reveal?

6. What does the presence of hornfels reveal?

GEOLOGIC TIME

Lesson 9

Purpose

The activities in this lesson will lay the foundation for understanding and applying techniques in interpreting the dating of igneous, sedimentary, and igneous rock layers. These techniques are used by geologists to determine the geologic history of an area.

Learning Objectives

After completing this laboratory lesson, you will be able to:

- Demonstrate the techniques used to determine the relative and absolute ages of rock strata.
- Explain how geologists using relative and absolute dating techniques as well as index fossils to reveal the geologic history of an area.
- Describe how the geologic time scale is organized.

Materials Needed

The activities will be performed using the following materials. Be sure you have all listed materials before starting the activity.

- ❏ Pencil
- ❏ Eraser
- ❏ Calculator

Overview

Today's geologists have learned how to not only read a specific rock sample or the rocks in a rock body, as described in earlier chapters, but how to interpret layers of igneous, metamorphic, and sedimentary rock across an area. In doing so, they reconstruct Earth's history, region by region.

A key activity in the interpretation of rock strata is dating the layers; that is, determining their ages. There are two types of dating, each of which yields a different type of information: relative dating and absolute dating.

Relative dating is a method of determining the age of an object relative to the age of another object. For example, based on physical appearance, one could easily conclude that a mother is older than her baby, who is younger than his big brothers and sisters. On the other hand, absolute dating determines an actual chronological age, perhaps that the mom is 32, whereas the baby is 6 months old.

Knowing the absolute ages of a sequence of events automatically gives information about relative age; however, this does not work in reverse. Knowing the sequence of events does not reveal information about the absolute age of those events. Even so, relative age is useful. For instance, in a culture where inheritance goes to the oldest child, we don't need to know the actual age of any of the family's children to know which child will get the farm. In the case of geology, relative age reveals the sequence of an area's history.

Relative Dating

Geologists apply a variety of principles to determine the relative age of rocks based their relationship to each other.

- The **Principle of Superposition** states simply that younger rocks are deposited over older rocks. This means that in a sequence of sedimentary rocks, unless overturned, the oldest beds are at the bottom and the youngest beds are at the top (**Figure 9.1**).

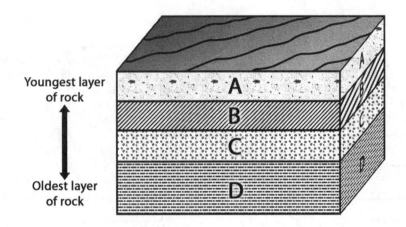

Figure 9.1 Principle of Superposition. The Principle of Superposition states that younger rocks are deposited over older rocks. Illustration by Bob Dixon.

- The **Principle of Original Horizontality** states that sedimentary layers of rock or lava flows accumulate as horizontal layers on an essentially horizontal surface, like sheets of paper in a book. This is best seen in photos of the Grand Canyon where sedimentary layers are visibly horizontal (**Figure 9.2**). If there is an area where sedimentary layers are tilted, the logical conclusion is that the tilting occurred after the rocks formed.

Figure 9.2 Principle of Original Horizontality. The fact that sedimentary deposits accumulate in horizontal layers is easily seen in Arizona's Grand Canyon. Photo: Shutterstock 9040255, credit Markrhiggins.

- The **Principle of Inclusion** refers to older rock fragments found within igneous intrusions. It states that the fragments, called inclusions, must be older than the rocks forming the intrusion in order to be surrounded by them. If they were part of the intrusion, they would have been part of the magma and not separate fragments; if they were younger than the intrusion, they could not have entered it because the igneous rock would already have solidified. (**Figure 9.3**).

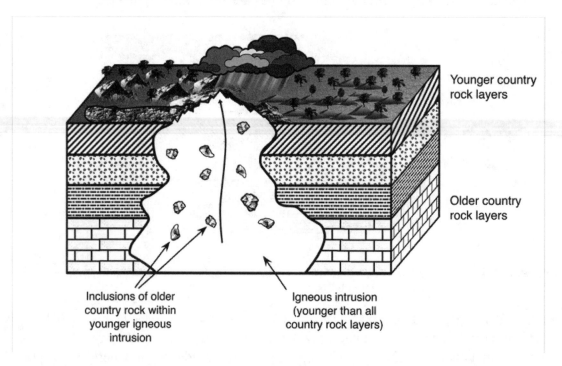

Younger country rock layers

Older country rock layers

Inclusions of older country rock within younger igneous intrusion

Igneous intrusion (younger than all country rock layers)

Figure 9.3 Principle of Inclusion. Rock fragments inside an igneous rock body are older than the igneous intrusion. Illustration by Bob Dixon.

- The **Principle of Lateral Continuity** states that a particular layer of sedimentary rock originally extended in all directions until the thickness either thinned to zero or until the layer reached the edge of the basin in which it was deposited. This principle allows geologists to recognize that a layer with an identical composition miles away may have been deposited at the same time and, therefore, under the same circumstances, helping them to reconstruct the history of a region.

- The **Principle of Cross-cutting Relationships** states that geologic features cutting across other geologic features are younger than the bodies they cut across (**Figure 9.4**). For example, an igneous intrusion cuts across country rock that was already there when the magma intruded, providing evidence that the intrusion is younger than the country rock. Cross-cutting features include fractures or cracks in rocks, faults, and dikes formed when magma was injected into cracks within country rock, then crystallized.

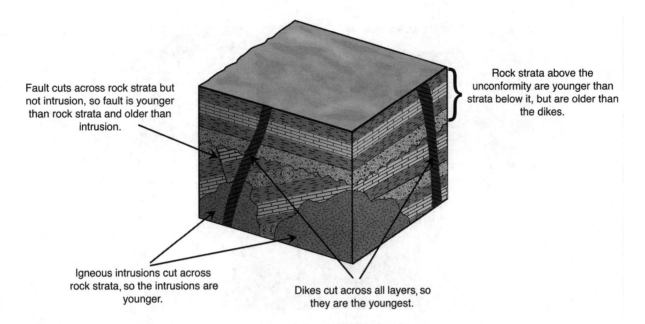

Fault cuts across rock strata but not intrusion, so fault is younger than rock strata and older than intrusion.

Rock strata above the unconformity are younger than strata below it, but are older than the dikes.

Igneous intrusions cut across rock strata, so the intrusions are younger.

Dikes cut across all layers, so they are the youngest.

Figure 9.4 Principles of Cross-cutting Relationships. A geologic structure is younger than any geologic structure that it cuts across. Illustration by John Renton.

- The **Principle of Faunal Succession** is sometimes called the Principle of Fossil Succession. It states that there is a unique, recognizable sequence of fossils in the rock record from the oldest to the youngest. Some individual species and groups of fossils, called index fossils, are characteristic of specific time frames in the past. Finding a rock with such fossils provides an immediate indication as to the age of the rock. This principle is based on observation and on the absolute dating methods described later in this lesson. It is the result of painstaking work finding, collecting, identifying, cataloging, and dating fossils found around the world.

Gaps in the Geologic Record

Reconstructing Earth's history would be simpler if rock layers were continuously deposited one after another. In fact, however, there are periods during which exposed rocks are weathered and carried off as sediment, erasing an interval in history, only to be later covered with a new layer of igneous or sedimentary rock. The erosional surface that separates the two rock sequences is called an unconformity; an unconformity represents a gap in time, called a hiatus, that no longer appears in the rock record.

There are three types of unconformity, each of which reveals a distinctive story about the rocks above and below the boundary: (1) angular unconformity, (2) disconformity, and (3) nonconformity.

- An *angular unconformity* occurs where horizontal layers of rock have been deposited over tilted layers, as shown in **Figure 9.5**. The Principle of Superposition tells us that the tilted, bottom layers are older, while the Principle of Original Horizontality tells us that these layers were horizontal when deposited. Thus, the angular unconformity reveals a story: (1) deposition and lithification of the bottom layers, followed by (2) tectonic stresses that tilted the layers, after which (3) a period of erosion removed parts of the tilted layers, and finally (4) new deposits were made that subsequently lithified.

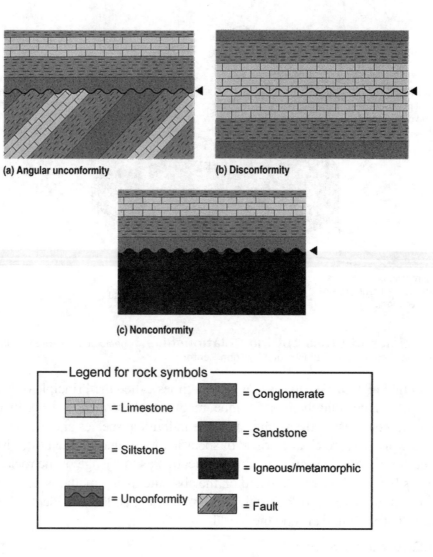

(a) Angular unconformity

(b) Disconformity

(c) Nonconformity

Legend for rock symbols

= Limestone

= Siltstone

= Unconformity

= Conglomerate

= Sandstone

= Igneous/metamorphic

= Fault

Figure 9.5 Types of Unconformities. From *Planet Earth* by John J. Renton. Copyright © 2002 by John J. Renton. Reprinted by permission of Kendall Hunt Publishing Company.

- A *nonconformity* is an unconformity formed when a layer of sedimentary rocks is deposited on the eroded surfaces of intrusive igneous rocks or metamorphic rocks. A nonconformity tells the following story: (1) intrusive or metamorphic rocks formed beneath the surface, then (2) the rocks were uplifted above the surface, where they and any overlying rocks were eroded, and finally (3) the surface subsided or the sea level rose to again submerge the rock, where younger sediments were subsequently deposited.

- A *disconformity* is the most difficult type of unconformity to detect. A disconformity is the boundary between two parallel beds of sedimentary rock. It's difficult to recognize because it's common to have two adjacent parallel beds of sedimentary rock with no unconformity between them. It's not always immediately obvious when a surface between two beds is actually a disconformity. One way a disconformity may form is as follows: (1) a layer or layers of sedimentary rock were deposited, then (2) the sea level dropped or the land was uplifted without local stresses that would have caused deformation, after which (3) the exposed rock layers were eroded, and then (4) the sea level rose or the land subsided to again submerge the eroded sedimentary layer, and finally (5) a new sedimentary layer was deposited and lithified.

Radiometric Dating

The absolute dating of rocks is largely accomplished through a technique called radiometric dating. Radiometric dating relies on a behavior of isotopes called radioactive decay. Isotopes are variations in the atoms of an element based on the number of neutrons the nucleus contains. The nucleus of an atom is composed of protons and neutrons. The number of protons is the same in all atoms of a given element; for instance, all carbon atoms contain six protons. However, the number of neutrons varies. Each variation is a different isotope. Isotopes of an element are identified by the total number of particles in a nucleus. For instance, the most common carbon nucleus has six protons and 6 neutrons, so it is called carbon-12 or ^{12}C (**Figure 9.6**).

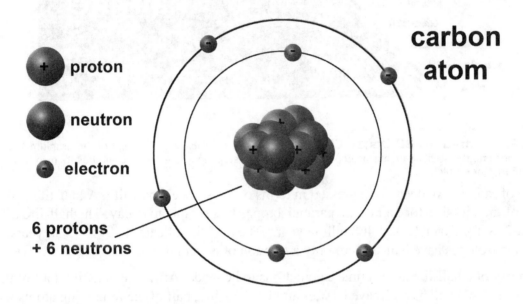

Figure 9.6 Carbon-12 Atom. All carbon-12 atoms have six protons and six neutrons, or 12 particles in their nuclei. By contrast, the carbon-14 isotope has six protons and eight neutrons, for a total of 14 particles. Illustration by Marie Hulett.

Many isotopes, particularly those of the heavier elements, are unstable; that is, their large nuclei have trouble holding together. Eventually, some of the particles will be emitted in a process called radioactive decay. Radioactive decay always involves a change in the number of protons in the nucleus. When the number of protons changes, the identity of the element changes because it is the number of protons that determines what element an atom is. The original isotope is called the parent isotope, and the new isotope is called the daughter isotope.

Decaying isotopes may go through a number of steps, called a decay chain, before they reach a form that is stable. For instance, the uranium isotope, uranium-238, goes through 14 steps, changing to a new element each time, before it achieves stability as lead-206 (**Figure 9.7**). This path takes time to accomplish; the exact amount of time varies with the isotope.

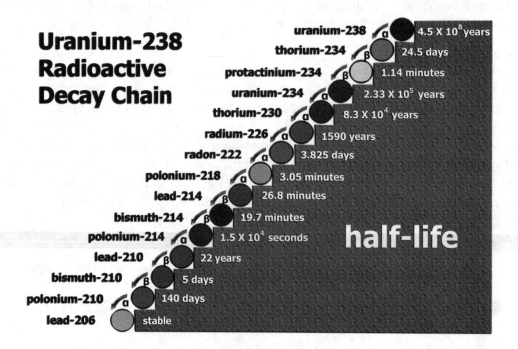

Figure 9.7 Uranium-238 Decay Chain. It takes 4.468 billion years for half of the uranium-238 in a sample to become the stable daughter isotope, lead-206. Each step along the decay chain has its own half-life. Illustration by Marie Hulett.

The rate of radioactive decay is expressed in terms of an isotope's half-life. A half-life is the amount of time it takes for half of the parent isotopes in a sample to decay. The half-life of uranium-238, for example, is 4.468 billion years. The half-life of carbon-14 is 5,730 years. Some radioactive isotopes have half-lives of only a fraction of a second.

The concept of a half-life allows time calculations to be made. After one half-life, half of the atoms in a sample will have decayed. After another half-life, half of the remaining atoms will have decayed—leaving behind only a quarter of the original number of atoms of the isotope. The chart in **Figure 9.8** shows how the proportion of the original isotopes dwindles as half-lives pass.

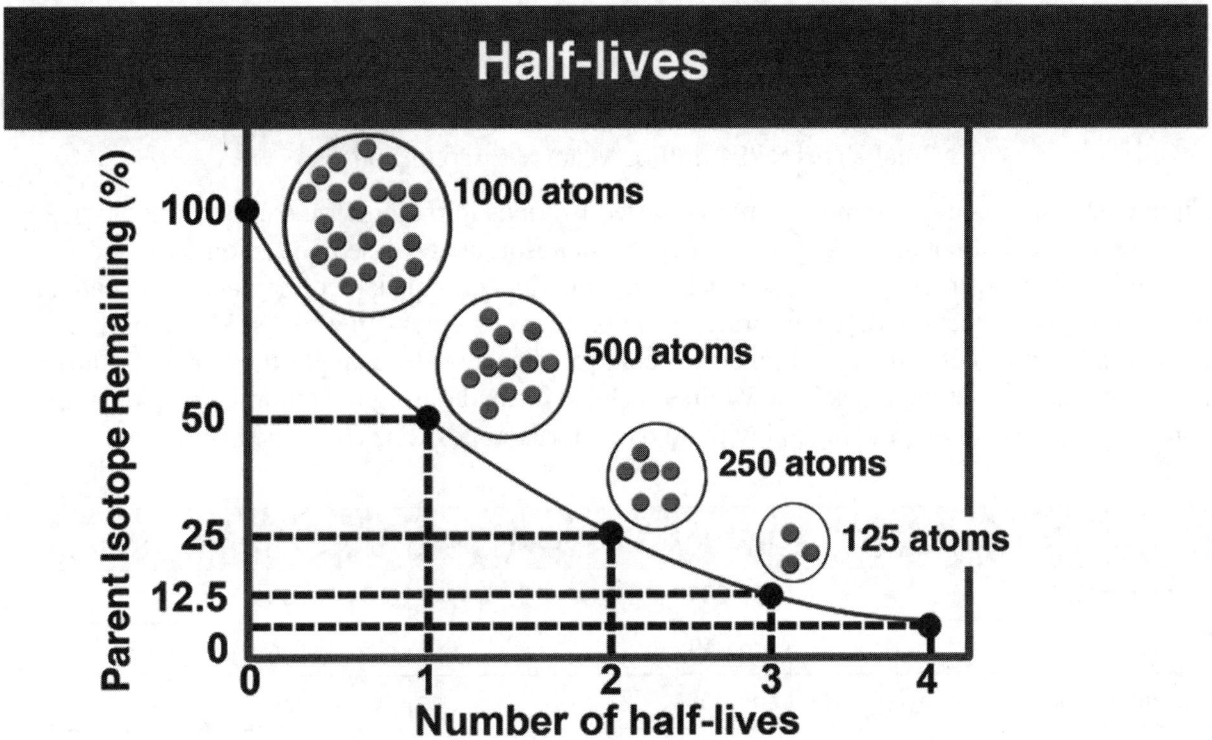

Half-lives

Figure 9.8 Half-Life. The half-life of a radioactive isotope is the amount of time required for one-half of the atoms of a parent isotope to disintegrate into atoms of the daughter isotope. Each half-life cuts the remaining amount of parent isotope in half. Illustration by Marie Hulett.

Radiometric dating is a method of assigning an absolute date to an object based on the rate of radioactive decay of a particular isotope. It is accomplished by:

(1) Using an instrument called a mass spectrometer to count the number of parent isotopes and daughter isotopes in a sample.

(2) Adding the number of parent and daughter isotopes together to determine the original number of parent isotopes in the sample, assuming that none of the isotopes has escaped since the rock was formed.

(3) Calculating the proportion of remaining parent isotopes to the original total.

(4) Determining the number of half-lives that have passed since the substance was formed based on the proportion in Step 3.

(5) Doing the math to determine how much time it has taken for the current proportion of isotopes in the sample to be reached.

For example, uranium-235 has a half-life of 704 million years. If the measurements from a mass spectrometer determined that the number of uranium-235 isotopes remaining in a sample of an igneous rock is 50 percent of the original number, that would mean that one half-life had passed since the original granitic magma cooled and solidified, and the estimated age of the rock would be 704 million years.

If it were determined that only 25 percent of the original number of uranium-235 isotopes remained, that would mean two half-lives had passed since the rock crystallized, and the calculated age of the rock would be:

2 half-lives × 704 million years = 1.426 billion years

Figure 9.9 lists some of the most commonly used isotopes in radiometric dating for geological purposes. Note that not all rocks can be assigned an absolute age based on radiometric techniques. Igneous rocks can be dated used radiometric techniques. Sedimentary rocks cannot be dated this way because they are formed from particles of preexisting rocks. Metamorphic rocks can often be dated using radiometric techniques because the changes they undergo during metamorphism "reset the clock." New minerals have formed during metamorphism and so the parent isotopes are now in a mineral with no daughter isotopes. The clock then starts again.

Parent Isotope	Daughter Isotope	Half-Life in Years
Uranium-235	Lead-207	704 million
Potassium-40	Argon-40	1.25 billion
Uranium-238	Lead-206	4.47 billion
Rubidium-87	Strontium-87	48.8 billion

Figure 9.9 Useful Elements for Radiometric Dating. Illustration by Susan Wilcox.

Index Fossils

As noted above, sedimentary rock cannot be assigned an age based on radiometric dating. However, geologists have found a couple of ways to date sedimentary rocks. One way is to find the absolute age of any adjacent igneous or metamorphic rocks, and then use relative dating to determine an approximate age range for the sedimentary layer. For instance, if an underlying metamorphic layer is 250 million years old and an overlying igneous layer is 200 million years old, then the Principles of Superposition tell us that a sedimentary layer between them must be between 200 and 250 million years old.

Another way to determine the age of a sedimentary layer is based on the fossil record. As noted earlier, the time frames associated with various index fossils are the result of painstaking work by many scientists. Since fossils form at the same time as a sedimentary layer, determining the age of a layer by dating adjacent igneous or metamorphic rocks also provides an age for the fossils. When a layer containing the same types of fossils is found elsewhere, it follows that the new layer will be approximately the same age as other layers containing those fossils. Wherever index fossils are found, they immediately provide a valid age for the rocks in which they are contained.

Earth's history as revealed by the fossil record has led to the development of a geologic time scale (**Figure 9.10**). It divides Earth's history into eons, eras, periods, and epochs based changes in the types of organisms whose fossils appear in sedimentary layers. As noted earlier, the Principle of Faunal Succession holds that there is a specific sequence of fossils that appear in the rock record. That sequence has allowed geologists to divide geologic history into time frames based on the types of organisms that left fossils behind. **Figure 9.10** reveals not only the organization of geologic time but also some of the organisms that lived in each time frame. This chart is based on the 2009 Geologic Time Scale published by the Geological Society of America. As scientists continually discover and analyze more data, the result is that they often make new estimates, usually after much discussion, so it is likely that you will encounter other geologic time scales that use slightly different numbers.

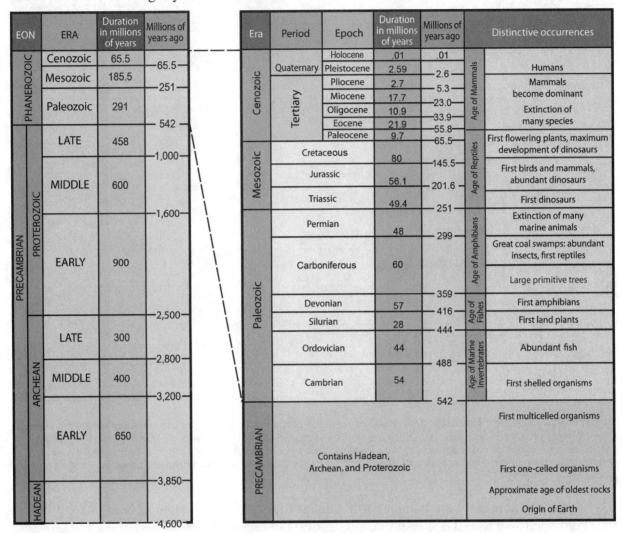

Figure 9.10 Geologic Time Scale. The geologic time scale is largely based on the chronological evolution of life as recorded in the fossils contained in sedimentary rocks. Illustration by John J. Renton.

Figure 9.11 shows some of the index fossils scientists commonly use to date sedimentary rocks. There are other fossils in the same rocks that are characteristic of the times—dinosaurs, small mammals, fish, reptiles—that can be used for the same purpose.

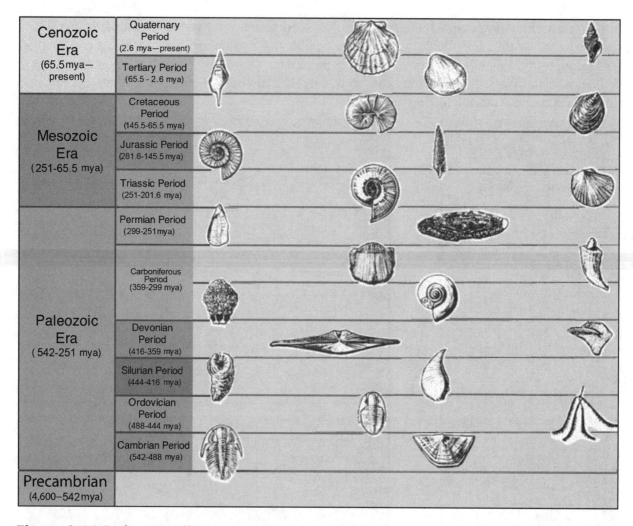

Cenozoic Era (65.5 mya–present)	Quaternary Period (2.6 mya—present)		
	Tertiary Period (65.5 - 2.6 mya)		
Mesozoic Era (251-65.5 mya)	Cretaceous Period (145.5-65.5 mya)		
	Jurassic Period (281.6-145.5 mya)		
	Triassic Period (251-201.6 mya)		
Paleozoic Era (542-251 mya)	Permian Period (299-251 mya)		
	Carboniferous Period (359-299 mya)		
	Devonian Period (416-359 mya)		
	Silurian Period (444-416 mya)		
	Ordovician Period (488-444 mya)		
	Cambrian Period (542-488 mya)		
Precambrian (4,600–542 mya)			

Figure 9.11 Index Fossils. This chart shows some of the fossils geologists use as index fossils for various time frames in the geologic time scale. Each index fossil only appeared during the time frame associated with it, measured as millions of years ago (mya). Consequently, when an index fossil is found in a sedimentary rock, it provides the time frame for that rock's formation. Chart by Don Vierstra and Marie Hulett, based on image courtesy of USGS.

It's sobering to note that even though Earth's oldest rocks have been dated as being 4.6 billion years old, the fossil record documenting an explosion of increasingly more complex life forms began only 542 million years ago—although there are some traces of simple organisms that go back as far as 3.5 billion years. This means that the evolutionary history revealed by the fossil record occupies only the latest one-ninth of Earth's history. More than four billion years elapsed between the birth of our planet and the time when life more complex than a single cell began to flourish and diversify, a time span known as Precambrian. The 10,000 years of recorded human history in the Holocene epoch is barely the blink of an eye compared to the vast expanse of geologic time.

Lab Exercise

Lab Exercise: *Dating Rock Strata*

During this lab exercise, you are going to determine the history of the geologic cross section in **Figure 9.12** through three methods:

    (1) relative dating of the geologic features;

    (2) absolute dating of the igneous and metamorphic rock layers;

    (3) the use of index fossils to determine the age of sedimentary layers.

Instructions

Step 1: List each of the geologic features shown in **Figure 9.12** from youngest (1) to oldest (8). Unconformities (i.e., angular unconformity, disconformity, and nonconformity) are indicated by a wiggly line between rock layers.

               Rock units A, D, E, and F are sedimentary.

               Rock units B and C are igneous.

               Rock unit G is metamorphic.

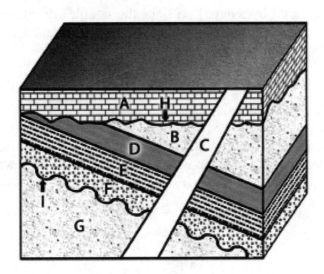

Figure 9.12 Geologic Cross Section of Relative Age Analysis. Illustration by Bob Dixon.

    1. _____ (youngest rock unit)

    2. _____

    3. _____

    4. _____

    5. _____

    6. _____

    7. _____

    8. _____

    9. _____ (oldest rock unit)

   10. What letters represents unconformities, and what type of unconformities are they?

Step 2: Calculate the ages of the metamorphic and igneous layers based on the following information.

- A mineral sample from rock unit B has 150,000 atoms of uranium-235 and 50,000 atoms of lead-207.
- A mineral sample from rock unit C has 180,000 atoms of uranium-235 and 20,000 atoms of lead-207.
- A mineral sample from rock unit G has 100,000 atoms of potassium-40 and 100,000 atoms of argon-40.

Use **Figure 9.8**, if necessary, to help you determine calculate how many half-lives have passed since rock units B, C, and G were each formed, based on the remaining amount of a parent isotope compared to the original amount of parent isotope.

Recall that the original amount of parent isotope can be calculated by adding together the amount of parent isotope plus the amount of daughter isotope, making the assumption that neither has escaped the sample. Also note that if less time has passed than a full half-life, you must calculate what portion of a half-life has passed to make the calculations.

Once you know how many half-lives have past since the sample crystallized, use the half-life table in **Figure 9.9** to calculate an absolute date in mya (million years ago) for each rock unit.

11. The igneous rock in rock unit B was formed _____.

12. The igneous rock in rock unit C was formed _____.

13. The metamorphic rock in rock unit G was formed _____.

Step 3: Index fossils have been found in the sedimentary layers! Use **Figure 9.11** to help you to identify the periods in which layers A, D, E, and F were formed as well as their age in mya based on the discovery of the following fossils:

14. The following fossils were found in Layer A:

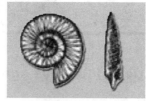

Figure 9.13

Layer A was formed _____ mya during the _____ Period.

15. The following fossils were found in Layer D:

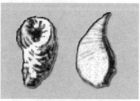

Figure 9.14

Layer D was formed _____ mya during the _____ Period.

16. The following fossils were found in Layer E:

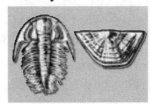

Figure 9.15

Layer E was formed _____ mya during the _____ Period.

17. The following fossils were found in Layer F:

Figure 9.16

Layer F was formed approximately _____ mya during the _____ Period.

18. What is the least amount of time that could have passed during the hiatus represented by youngest unconformity seen in **Figure 9.12**?

19. What is the least amount of time that could have passed during the hiatus represented by the older unconformity seen in **Figure 9.12**?

Online Activities

Per your instructor's directions, go to the online supplement for this lab and complete the activities assigned. Viewing the online videos will help you to complete the quiz.

Quiz

Multiple Choice

Questions 1 through 3 are based on the **Lab Exercise, Step 2.**

1. Record your answer to **Lab Exercise, Step 2, Question 11.** How long ago was the igneous rock in rock unit B formed?

   a. 352 mya

   b. 380 mya

   c. 704 mya

   d. 1408 mya

2. Record your answer to **Lab Exercise, Step 2, Question 12.** How long ago was the igneous rock in rock unit C formed?

   a. 70.4 mya

   b. 140.8 mya

   c. 352 mya

   d. 704 mya

3. Record your answer to **Lab Exercise, Step 2, Question 13.** How long ago was the metamorphic rock in rock unit G formed?

   a. 625 mya

   b. 704 mya

   c. 1250 mya

   d. 2500 mya

Questions 4 through 8 are based on the **Lab Exercise, Step 3.**

4. Record your answer to **Lab Exercise, Step 3, Question 14.** How long ago were the fossils found in Layer A formed?

   a. 65.5–145.5 mya

   b. 201.6–145.5 mya

   c. 251–201.6 mya

   d. 299–251 mya

5. During what period were the fossils in Layer A formed?

   a. Cretaceous Period

   b. Triassic Period

   c. Jurassic Period

   d. Cambrian Period

6. Record your answer to **Lab Exercise, Step 3, Question 15.** How long ago were the fossils found in Layer D formed?

   a. 251–299 mya

   b. 359–299 mya

   c. 416–359 mya

   d. 444–416 mya

7. During what period were the fossils in Layer D formed?

   a. Permian Period

   b. Devonian Period

   c. Silurian Period

   d. Ordovician Period

8. Record your answer to **Lab Exercise, Step 3, Question 16**. How long ago were the fossils found in Layer E formed?

   a. 444–416 mya

   b. 488–444 mya

   c. 542–488 mya

   d. 1000–542 mya

9. During what period were the fossils in Layer E formed?

   a. Permian Period

   b. Devonian Period

   c. Silurian Period

   d. Ordovician Period

10. Record your answer to **Lab Exercise, Step 3, Question 17**. How long ago were the fossils found in Layer F formed?

    a. 444–416 mya

    b. 488–444 mya

    c. 542–488 mya

    d. 1000–542 mya

11. During what period were the fossils in Layer F formed?

    a. Cambrian Period

    b. Triassic Period

    c. Carboniferous Period

    d. Tertiary Period

12. Record your answer to **Lab Exercise, Step 3, Question 18**. What is the least amount of time that could have passed during the hiatus represented by younger unconformity seen in **Figure 9.12**?

    a. 101 million years

    b. 150.4 million years

    c. 206.6 million years

    d. 286 million years

13. Record your answer to **Lab Exercise, Step 3, Question 19**. What is the least amount of time that could have passed during the hiatus represented by the older unconformity seen in **Figure 9.12**?

    a. 708 million years

    b. 762 million years

    c. 1333 million years

    d. 1958 million years

14. Which of the following principles is **NOT** used in determining the relative age of rocks?
    a. Principle of Cross-cutting Relationships
    b. Principle of Original Horizontality
    c. Principle of Superposition
    d. Principle of Vertical Continuity

Determine the relative age each of the geologic structures shown in the geologic cross section below (**Figure 9.17**) from youngest to oldest; then answer the questions below. As before, a wiggly line running across the geologic cross section indicates an unconformity and a fault is indicated with a bold straight line.

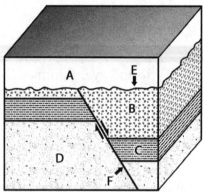

Figure 9.17 Geologic Cross Section of Relative Age Analysis. Illustration by Bob Dixon.

15. Which is the youngest geologic structure in **Figure 9.17**?
    a. E
    b. A
    c. B
    d. D

16. Which is the oldest geologic structure in **Figure 9.17**?
    a. A
    b. C
    c. F
    d. D

17. Which letter represents an unconformity in **Figure 9.17**?
    a. A
    b. C
    c. E
    d. F

18. Which letter represents a fault in **Figure 9.17**?
    a. A
    b. C
    c. E
    d. F

19. What is the proper order of geologic structures B, C, D, and F in **Figure 9.17** from oldest to youngest?
    a. F, B, C, D
    b. D, C, B, F
    c. D, F, B, C
    d. B, C, F, D

20. What relative dating principle states that in a sequence of sedimentary rocks, unless overturned, the oldest beds are at the bottom and the youngest beds are at the top?
    a. Principle of Cross-cutting Relationships
    b. Principle of Inclusion
    c. Principle of Superposition
    d. Principle of Original Horizontality

21. What relative dating principle states that fossils in a sequence of sedimentary rock layers succeed one another in a definite, recognizable order?
    a. Principle of Cross-cutting Relationships
    b. Principle of Faunal Succession
    c. Principle of Inclusion
    d. Principle of Superposition

22. The Principle of Cross-cutting Relationships states that
    a. sedimentary layers and lava flows accumulate as horizontal layers.
    b. rock fragments found within an igneous intrusion are older than the intrusion.
    c. geologic structures that cut across other structures are younger than the structures they cut across.
    d. younger rocks are deposited over older rocks.

23. The Principle of Lateral Continuity states
    a. sedimentary layers and lava flows accumulate as horizontal layers.
    b. a layer of sedimentary rock originally extended in all directions until it thinned to zero or reached the edge of the basin.
    c. rock fragments found within an igneous intrusion are older than the intrusion.
    d. younger rocks are deposited over older rocks.

24. What type of unconformity appears between two parallel sedimentary layers?
    a. a disconformity
    b. a nonconformity
    c. an angular unconformity
    d. a sedimentary nonconformity

25. What type of unconformity appears between a sedimentary layer and an igneous or metamorphic rock body?
    a. a disconformity
    b. a nonconformity
    c. an angular unconformity
    d. a sedimentary nonconformity

26. What type of unconformity appears between a tilted sedimentary strata and a horizontal overlying layer?

    a. a disconformity

    b. a nonconformity

    c. an angular unconformity

    d. a sedimentary nonconformity

27. If a sample of radioactive material contains a parent isotope with a half-life of 3 years, then at the end of 6 years

    a. all of the parent isotope remains.

    b. half of the parent isotope remains.

    c. one-quarter of the parent isotope remains.

    d. one-ninth of the parent isotope remains.

Short Answer
Questions 1 and 2 are based on the **Lab Exercise, Step 1.**

1. Record your answers to **Lab Exercise, Step 1, Questions 1 through 9**. List each of the geologic features shown in **Figure 9.12** from youngest to oldest.

    _____ (youngest rock unit)

    _____

    _____

    _____

    _____

    _____

    _____

    _____

    _____ (oldest rock unit)

2. Record your answer to **Lab Exercise, Step 1, Question 10**. What letters represents unconformities, and what type of unconformities are they?

3. A _____ is the preserved remains or traces of a once-living organism.

4. An _____ is a fossil of an organism that was common and had widespread geographic distribution during a short period of time in Earth's history.

5. Why is radiometric dating not used to determine the age of sedimentary rocks?

6. How have fossils contributed to the organization of the geologic time scale?

MASS WASTING

Lesson 10

Purpose

The activities in this lesson will lay the foundation for understanding and applying techniques in interpreting slope stability and how landslides are created. Each year, thousands of individuals are displaced by landslides that damage or destroy physical structures (i.e., homes, businesses, and property) and cost the individual taxpayer, insurance companies, and government organizations millions of dollars per year. You will learn how the mass movement of loose sediment, soils, and rocks down a slope occurs and some indications that a slope may fail.

Learning Objectives

After completing this laboratory lesson, you will be able to:

- Describe the major factors involved in the movement of materials downslope.
- Differentiate among the common types of mass wasting.
- Describe some indications of slope instability and ongoing mass movement.

Materials Needed

- ❏ Pencil
- ❏ Eraser

Overview

Mass wasting, the downslope movement of earth materials due to gravity, has been called the unsung hero of all geologic processes. Most individuals are aware of the power of erosion to wear down a hillside, but mass wasting tends to take them by surprise—generally when it is of the sudden, devastating, and unheroic variety, such as the landslide that destroys a village. While such an event is considered a natural disaster as a result of its impact on human beings, more subtle mass wasting is occurring continuously on almost all slopes. The rocks and soils on Earth's slopes follow the time-honored physical law: what goes up must come down. Sometimes the downward movement is quick and sudden, other times it is slow and relentless. Oftentimes, mass wasting is the first step in the erosion of a landscape. Streams and rivers mostly transport material that has been delivered to them by some form of mass wasting. In any case, earth materials move downward, forever changing the landscape.

Gravity and Resisting Factors

Every slope, no matter how gentle or steep, is subjected to the driving force of gravity that constantly pulls the earth materials composing it downward. Whether the materials succumb to the pull of gravity depends on the presence of resisting factors. There are three resisting factors at work on a slope: (1) friction, (2) cohesion, and (3) the strength of the earth materials.

- Friction is the resistance that occurs when one object slides over another. Friction among particles in slope materials helps to hold them in place and must be overcome before any substance can begin to move.

- Cohesion is the tendency for like particles to stick together. For instance, the particles in mud can be cohesive, which is why it sticks together during a ride on your shoes after a rainstorm. Dry sand, on the other hand, lacks cohesion so the particles don't stick together unless cemented together.

- Strength is a material's resistance to stress. A strong rock, like granite, will not easily break apart, whereas a weak rock, like shale, will fracture under far less stress. The strength of slope materials plays a particularly important role when a slope is composed of rock layers that dip parallel to the slope. The downward pull of gravity places a tensional stress on rock beds; if the rock composing the beds is weak, it will respond with brittle stress; that is, it will fracture and move downslope carrying overlying layers of rock, soil, and vegetation with it.

Mass Wasting Triggers

When rock layers underlying the surface of a slope are strong and cohesive, they don't easily succumb to the force of gravity. In this case, the slope is stable, or resistant to downward movement. If the rock layers are not cohesive or if they are weak, gravity may be strong enough to overcome the forces holding them together. Such slopes are referred to as unstable because they are likely to succumb to the force of gravity, sometimes quite suddenly.

The angle of the slope at which a balance between gravity and resisting factors is typically achieved is called the angle of repose. The angle of repose varies depending on the slope materials. For dry sand, the angle of repose is about 45 degrees. More angular particles have a steeper angle of repose than rounder particles. If the angle of repose is exceeded for some reason, the materials will begin to slide and move downslope until a stable angle is achieved.

In general, most slopes exist day after day with the force of gravity balanced by resisting factors. However, other factors, called triggers, can upset this balance. Common triggers are earthquakes and other vibrations, an excess of water, an increase in weight placing pressure on the slope's surface, and undercutting.

Earthquakes and other vibrations are triggers because the shaking can disrupt the cohesion between particles. Once cohesion is reduced, the force of gravity may be enough for slope materials to come tumbling downward. In the western United States, a major earthquake without some mass wasting is a relatively rare event.

Mass wasting is also common after major rainstorms or snowmelts. During a rainstorm or snowmelt, water permeates the soil and sediments until they are saturated, meaning that every particle is surrounded by water. The water is a lubricant, which, by definition, is a substance introduced between two moving surfaces to reduce friction and cohesion. Once cohesion between particles is lost, the particles succumb to gravity, sliding or flowing down the slope. Or the water may get in between rock layers, lubricating their surfaces until they no longer stick together, leading to the upper layer sliding down over the lower layer. Such events can happen weeks after a major storm as the water permeates deeper into soil and other earth materials.

Increased weight can cause a formerly stable slope to become unstable; for instance, when there is construction at the top of the slope, or if saplings on a slope grow into towering trees. The weight magnifies the effect of gravity by placing additional downward pressure on the slope. The combination may be enough to overcome resisting forces, especially when another trigger is present. Water plays a role here as well; in addition to its role as a lubricant, water added to slope materials tends to increase their total mass, which increases the force exerted by gravity.

At the same time, the roots of those towering trees provide extra friction and help to hold soil in place. The impact of vegetation on a slope's stability is complex and depends on the extent and types of plant cover. In general, the root system of a continuous plant cover promotes slope stability by increasing friction and cohesion; however, a continuous plant cover can also decrease slope stability not only by adding weight but also by reducing runoff and increasing the amount of water that soaks into the soil—even though the plants themselves will extract some of the water for their own use, which then decreases the water in the soil. The impact of vegetation on slope stability depends on so many factors that there is no one rule of thumb for its use.

A final trigger is undercutting, the removal of lower portions of a slope. Undercutting is a common occurrence along rocky shorelines where waves erode the underlying portions of cliffs. Human beings are another cause of undercutting. The construction of roads in hilly or mountainous areas or of developments at the base of hills or mountains frequently undercuts the slopes. Removing support from the bottom of a slope is almost certain to eventually cause slope movement.

Types of Mass Wasting

There are many different types of mass wasting as well as many different ways to classify them. Part of the difficulty is that one type of movement can turn into another if the angle of the slope changes or the movement triggers a new movement. In addition, Mother Nature frequently refuses to accommodate categories, often providing conditions that seem to fall between two categories.

The classification scheme used here focuses on the direction of movement and the nature of the movement as ways to sort mass wasting events into categories. The chart in **Figure 10.1** summarizes this classification.

Type	Subtype	Movement	Angle	Materials
Fall	Rockfall	Rapid, individual units	Vertical	Rock fragments
	Avalanche	Rapid, picking up additional material along the way	Near-vertical	Snow, ice, rock, debris
Landslides	Landslide	Rapid, mass movement	Diagonal, along a plane parallel to slope	Rock, soil, regolith, plants
	Rock slide	Rapid, mass movement	Diagonal along a plane parallel to slope	Rock, soil
	Debris slide	Rapid, mass movement	Diagonal along a plane parallel to slope	Rock, soil, debris
	Slump	Rapid or slow, mass movement	Rotational, curving diagonally	Soil, regolith, plants
Flow	Mudflow	Rapid, flowing	Varies, follows stream valleys	Water, soil, regolith
	Lahar	Rapid, flowing	Varies, follows stream valleys	Water, ash
	Debris flow	Rapid, flowing	Varies, follows stream valleys	Water, soil, regolith, debris
	Earthflow	Moderate	Diagonal, along slope	Water, soil

Figure 10.1 Types of Mass Wasting. Illustration by Don Vierstra.

Falls

A **fall** involves rapid movement at a steep, nearly vertical angle. The most common types of falls are rockfalls and avalanches.

Rockfalls

The most rapid process of mass wasting is the rockfall where masses of rock free fall from vertical or near-vertical slopes (**Figure 10.2**).

Figure 10.2 Rockfall. Photo: Shutterstock 939535, credit GCRO Images.

Rockfalls may involve rocks of any dimension. Typically, rocks exposed at the surface of steep cliffs or roadcuts are subjected to various processes of weathering that slowly break them down along natural fractures. In addition, water that freezes in cracks and joints expands to break the rocks apart in a process called frost wedging that gradually dislodges them from the faces of steep outcrops. Eventually, various processes of erosion may undercut the cliff or roadcut, or the rock mass may be subjected to a local shock, such as an earthquake. Either occurrence can be a trigger for a rockfall.

Avalanche

While most people associate avalanches with snow, the term actually applies to a type of movement rather than the material involved. Avalanches occur on very steep slopes rather than vertical slopes. During an avalanche, falling rocks knock loose rocks below them; these, in turn, knock loose other rocks even farther down, and so on, until a massive pile of rocks, soil, and/or debris tumbles down the slope together. (**Figure 10.3**)

A snow avalanche occurs when the weight of accumulated snow and ice exceeds the strength and cohesion of the layers. Sometimes the balance between gravity and friction is so precarious that the smallest vibrations, such as the sound of a passing truck, can tip the scales in gravity's favor. Then the snow and ice tumbles down the slope, engulfing whatever is in its path.

Figure 10.3 Avalanche. Photo: Shutterstock 44773900, credit Kapu.

Landslides

Landslides involve the movement of soil and rock layers down gentler slopes than those of rockfalls or avalanches. *Landslide* is actually a collective term for a movement of a coherent mass of slope materials along what is more or less a plane. In most cases, the movement is similar to that of a child moving down a playground slide: straight with no rotation (**Figure 10.4**).

Figure 10.4 Landslide. A landslide occurs when a slope fails near the top and underlying materials slide down a plane. Photo: Shutterstock 36569911, credit Vadim Ostrikov.

As with the playground slide, the nature of the surface is one that lessens the effectiveness of friction. Slide planes are often associated with shale or slate rock layers. Because these rocks are impervious to water, water can collect on their surface to act as a lubricant for overlying layers. Shale and slate themselves are composed of clay minerals, which become slippery when wet, providing further lubrication. Finally, slate and shale have slaty cleavage that provides planes of weakness along which the rock layer can break.

Given enough weight, enough water, or the presence of other triggers like the shaking of an earthquake, such a slope will fail and the rock layers will slide down as a unit, carrying soil, debris, plants, and even trees with them.

Often geologists distinguish among landslides by naming them after the materials involved. A rock slide is a landslide composed mostly of loose rocks, for example, and a debris slide is one in which the slope materials include logs and brush.

A **slump** is a particular type of landslide during which the moving materials rotate as they slowly slip down the slope, creating a curved depression in the hillside. This type of motion occurs when the slope failure is near the bottom of the slope. As the lower part of the slope moves, it removes support from the top of the slope, and everything above the failure slides down. Think of the way a teenager slumps in a chair. As his lower back releases support, his spine curves and he slides down the chair, his feet moving forward. The slope materials in a slump move in the same, curved way.

Slumps are identifiable by their appearance. The rotational movement often leaves behind a curved scarp at the top of the slump; below that, the slope has a stepped appearance (**Figure 10.5**).

Figure 10.5 Slump. A slump is a slow downward, rotational movement of material that occurs when the support below a slope is removed, and is recognizable by its curved scarp and stepped appearance. Illustration by John J. Renton; Photo: Courtesy of USGS.

Flows

Flows are distinguished from slides in that the downward moving material in a flow has mixed with water to form a mud or a slurry. There are several circumstances under which this can occur, creating mudflows, earthflows, debris flows, and the much slower movements known as solifluction and creep.

Mudflows

Most mass wasting processes carry materials only a short distance, a few hundred feet in most cases. By contrast, mudflows are capable of transporting materials for many miles. Most mudflows consist of relatively small particles, mostly silts, clays, and ashes, carried in large volumes of water. However, because they are so viscous, mudflows may pick up larger particles as they flow, including cobbles, boulders, even houses and cars. They generally follow existing stream channels.

Debris Flows

Debris flows are mudflows that carry larger particles—sand-sized or larger—as well as debris, such as logs, branches, and brush (**Figure 10.6**). The origins of mudflows and debris flows are similar in that both are usually the result of torrential rains or snowmelts that soak unstable slopes composed of weathered rock particles and/or soil. A common preexisting condition for a debris flow is a slope that has been deforested by wildfire. Slopes in the burn area lack protection for the soil, and are likely to fail with the first heavy rain, carrying slope materials and burnt debris in the flow. Debris flows can travel as slow as a few feet per year or as fast as 320 kilometers per hour (200 miles per hour), depending on the angle of the slope and on the mass and viscosity of the materials.

Figure 10.6 Debris Flow. Debris flows are especially common after wildfires have denuded slopes. The flows are likely to carry burnt debris as well as the larger particles typical of debris flows. Photo: Courtesy of USGS.

One dramatic example of a mudflow or debris flow (depending on exact composition) is the lahar, a type of flow that occurs as the result of an explosive volcanic eruption. During a lahar, the heat of the eruption melts the snowcap that typically covers stratovolcanoes. The ejected ash and mud from the eruption then mix with the water to form a mud that travels rapidly, usually—but not always—down stream valleys. The largest recorded lahars in the United States have been the result of volcanic activity in the Cascade Mountains of the Pacific Northwest, such as the 1981 eruption of Mount St. Helens. That eruption produced a lahar that rushed down the volcano's slopes at speeds as high as 50 miles per hour near the top and

3 miles per hour at the bottom (**Figure 10.7**). Some witnesses reported a 12-foot-high wall of muddy water and debris.

Figure 10.7 Lahar. The lahar, a mudflow composed of volcanic ash and debris, traveled at speeds as high as 50 miles per hour during the 1982 eruption of Mount St. Helens in Washington State. Photo: Courtesy of Tom Casadevall/USGS.

Besides their role in volcanic eruptions, mudflows are especially common in arid and semiarid regions where there is little plant cover. Here, an intense rainfall runs off rather than soaks into the soil, picking up particles as it flows downslope until there is enough material within it to classify it as a mudflow. Similarly, mudflows are also common in regions that have recently been denuded of vegetation, such as in clear-cut timbering operations.

Earthflow
An earthflow is a flow of loose materials and water that occurs under a layer of sod. In temperate regions, slopes covered with sod become soaked during the rainy seasons of spring and early summer. The large amount of water mixing with earth materials under the sod decreases the cohesion and friction between the particles that normally stabilize the slope. Gravity then does its work, and the wet mixture moves downhill, but not always with enough force to tear the sod layer. If it does tear the sod layer, it is likely to turn into a flow or a slump, depending on the amount of water involved.

Solifluction
At higher latitudes and elevations where the ground is frozen throughout the year, except for a very short summer season, a type of flow called solifluction takes place when the uppermost layers of earth materials thaw (**Figure 10.8**). Because the earth materials below the thawed layer remain permanently frozen, the water accumulates and saturates the thawed layer. This process essentially eliminates cohesion and friction between the individual particles, literally converting the materials into a viscous liquid. With no force to hold the materials in place, the

surface materials slowly move downslope beneath the overlying vegetation at rates ranging from 0.5 centimeters to 5 centimeters per year (0.25 to 2 inches) on slopes angles as low as 2 to 3 degrees.

Figure 10.8 Solifluction. As earth materials between surface vegetation and the underlying layers of permafrost become saturated with water, the material will literally liquefy and form lobes of actively moving layers. These begin to move on very shallow slopes by the process called solifluction. Photo: © Steve Maslowski/Visuals Unlimited, Inc.

Creep

Creep is most subtle of the mass wasting processes. Creep is the imperceptibly slow, downslope movement of the upper layers of earth materials. Of all the processes of mass wasting, creep is the most widespread and is responsible for the movement of the greatest mass of material. As such, creep epitomizes the description of mass wasting processes being the unsung heroes of geologic processes. Because of creep's slow movement, the average individual is totally unaware of its existence.

Creep is the result of cycles of freezing and thawing or wetness and drying. The cycle begins when the soil becomes wet or when it freezes; either event lifts the particle perpendicular to the slope. When the soil dries or thaws, the particle settles ever so slightly downhill of its original location. In this way, the particles move down the slope a little with every cycle (**Figure 10.9**).

Figure 10.9 Creep. Left: Creep occurs from cycles of freezing and thawing or wetness and drying that gradually move particles downhill. Right: Bowed trees are evidence of creep. Illustration: From *Planet Earth* by John J. Renton. Copyright © 2002 by John J. Renton. Reprinted by permission of Kendall Hunt Publishing Company. Photo: Courtesy of Susan Wilcox.

Although the process occurs too slowly to actually observe, evidence for creep can be seen in tilted tombstones and fence posts, bent tree trunks, and the downslope bending of layers of shale.

Indications of Slope Instability

There are many indications of slope instability. Some of the best are evidence that mass wasting has occurred in recent history, increasing the likelihood that it will occur again in the future. Evidence of past mass wasting and other indications of slope instability include the following:

- A scar on a slope. A landslide generally leaves behind a tongue-shaped area of bare soil or rock on a hillside. In the case of a slump, the scar will include a curved scarp and a stepped appearance.

- A hummocky surface at the base of a slope. The area where slope materials have come to rest may become overgrown with vegetation, which gives them a lumpy appearance known as hummocky. Hummocky surfaces may also signal the beginning of an earthflow flow that may eventually develop into a slump.

- Bowed trees, tilted gravestones, and the bending of visible rock layers. These are evidence that creep is occurring. Bowed trees particularly point to creep; they occur when young trees continue to grow vertically even as the creep causes them to tilt forward. When the trees grow large enough, they are able to maintain a vertical orientation, but the lower section of the trunk remains bowed.

- Tilted trees, poles, fence posts, or visible rock layers. Unlike the bend that occurs as the result of creep, the tilt referred to here extends from the top of the tree (or pole) to the surface. Because trees always grow vertically, any tilt indicates the movement of the slope around the tree's roots. Likewise, utility poles are vertical when installed, so tilted poles indicate some movement has occurred. Lines between utility poles that are tighter or looser than usual can also indicate a movement of the pole as a result of slope movement.

- Areas of undercutting. Because undercutting is one of the triggers for mass wasting, any area of undercutting is an indication that a future mass wasting event is nearly certain.

- Cracks in pavement and fissures in the ground. Most pavement has some cracks, but if the road is on a slope and the crack extends parallel to the slope, the crack may be an indication that the slope is already moving or about to fail. Likewise, fissures extending along a slope are likely to be created from earth movement.

- Homes built on areas of creep will show the slow separation of wall joints, cracking in plaster and mortar, and movement of doors out of alignment with the door frame.

Lab Exercise

Lab Exercise: *Downslope Movement*

In this lab exercise, you will examine eight photographs of areas that have either had a mass wasting event or might in the future. As you examine **Figures 10.10** through **10.16**, answer the questions about photograph you are seeing. Make sure to save your results and observations. You will need this data to complete the quiz.

Instructions

Step 1:   Observe **Figure 10.10**.

Figure 10.10 Lab Exercise Step 1. Photo: Shutterstock 1070199, credit Ilya D. Gridnev.

What indication do you see that the slope in **Figure 10.10** might be unstable?

Step 2:   Observe **Figure 10.11**.

Figure 10.11 Lab Exercise Step 2. Photo: Shutterstock 8542513, credit Dr. Morley Read.

What type of mass wasting has occurred in **Figure 10.11**? What is the evidence that leads you to draw this conclusion?

Step 3: Observe **Figure 10.12**.

Figure 10.12 Lab Exercise Step 3. Photo: Courtesy of Liz Roll/FEMA.

What type of mass wasting has occurred in **Figure 10.12**? What is the evidence that leads you to draw this conclusion?

Step 4:   Observe **Figure 10.13**.

Figure 10.13 Lab Exercise Step 4. Photo: Shutterstock 75414961, credit Thor Jorgen Udvang.

Is the slope at the bottom of **Figure 10.13** stable or unstable? Why or why not? If the slope fails, what kind of mass wasting is most likely to occur?

Step 5: Observe **Figure 10.14**.

Figure 10.14 Lab Exercise Step 5. Photo: Shutterstock 131957, credit John Waller.

What type of mass wasting is occurring in **Figure 10.14**? What is the evidence that leads you to draw this conclusion?

Step 6:   Observe **Figure 10.15**.

Figure 10.15 Lab Exercise Step 6. Photo: Shutterstock 81705820, credit Yuriy Kulyk.

What indication do you see that the slope in **Figure 10.15** might be unstable? What type of mass wasting would this situation produce?

Step 7: Observe **Figure 10.16**.

Figure 10.16 Lab Exercise Step 7. Photo: Shutterstock 125589, credit Gordon Swanson.

What type of mass wasting has occurred in **Figure 10.16**? What is the evidence that leads you to draw this conclusion?

Online Activities

Per your instructor's directions, go to the online supplement for this lab and complete the activities assigned. Viewing the online videos will help you to complete the quiz.

Quiz

Multiple Choice

Questions 1 through 15 are based on the **Lab Exercise**.

1. In **Lab Exercise, Step 1**, what indication is there that the slope in **Figure 10.10** might be unstable?
   a. Tilted trees are evidence of slope movement.
   b. Bowed trees are evidence of creep.
   c. None, this slope appears to be stable.
   d. The ground has a hummocky surface.

2. In **Lab Exercise, Step 2**, what type of movement has occurred in **Figure 10.11?**
   a. A free fall in a vertical direction has occurred.
   b. A slide down a plane has occurred.
   c. A slide that has rotated created a curved surface.
   d. A flow similar to that of a viscous fluid has occurred.

3. What types of particles were involved in the movement seen in **Figure 10.11**?
   a. large blocks of rock
   b. silt- and clay-sized particles only
   c. particles in a range of sizes along with some debris
   d. trees and large boulders

4. Based on what you can see in **Figure 10.11**, what type of mass wasting event was this?
   a. landslide
   b. slump
   c. mudflow
   d. debris flow

5. If you knew that the event in **Figure 10.11** was the result of a volcanic eruption, how would you classify it?
   a. mudflow
   b. lahar
   c. debris flow
   d. solifluction

6. In **Lab Exercise, Step 3**, what is the distinguishing feature of the mass wasting seen in **Figure 10.12?**
   a. It has a stepped appearance.
   b. It was a flow.
   c. It involved free-falling rock particles.
   d. It occurred under the vegetation.

7. What type of mass wasting event occurred in **Figure 10.12**?
    a. creep
    b. slump
    c. mudflow
    d. solifluction

8. In **Lab Exercise, Step 4**, what indication is there that the slope in **Figure 10.13** might be unstable?
    a. It has a stepped appearance.
    b. None. This slope appears to be stable.
    c. It is a recent burn area and no longer has vegetation to protect the slope.
    d. There are no visible rock strata.

9. Under what conditions is the slope in **Figure 10.13** likely to fail?
    a. A cold winter is likely to cause frost wedging leading to rock fall.
    b. None. This slope appears to be stable.
    c. A windstorm may add enough downslope force for failure to occur.
    d. A heavy rainfall is likely to cause a debris flow.

10. In **Lab Exercise, Step 5**, what type of movement is occurring in **Figure 10.14**?
    a. solifluction
    b. earthflow
    c. creep
    d. None, this slope appears to be stable.

11. What indication of slope movement is evident in **Figure 10.14**?
    a. fissures in the slope
    b. tilted fence poles
    c. hummocky ground
    d. None, this slope appears to be stable.

12. In **Lab Exercise, Step 6**, what indication is there that the slope in **Figure 10.15** might be unstable?
    a. Hummocky ground along the edge of the slope.
    b. Undercutting has removed support for the slope.
    c. There is a lack of vegetation along the slope.
    d. None, this slope appears to be stable.

13. If the slope in **Figure 10.15** fails, what type of mass wasting event is it likely to produce?
    a. rockfall
    b. avalanche
    c. landslide
    d. slump

14. In **Lab Exercise, Step 7**, what is the distinguishing feature of the mass wasting seen in **Figure 10.16**?
    a. It has a stepped appearance.
    b. It was a flow.
    c. The slope materials moved down along a plane.
    d. It occurred under the vegetation.

15. What type of mass wasting event occurred in **Figure 10.16**?
    a. slump
    b. landslide
    c. lahar
    d. rockfall

16. What is mass wasting?
    a. an accelerated form of erosion
    b. any movement of slope materials
    c. a form of weathering
    d. the downslope movement of earth materials due to gravity

17. One characteristic of mass wasting processes is that they
    a. move materials very slowly.
    b. move materials relatively short distances compared to streams.
    c. only operate on steep slopes.
    d. operate only during nonfreezing months of the year.

18. What is the driving force for mass wasting?
    a. weight
    b. vibrations
    c. gravity
    d. erosion

19. Which of the below is **NOT** a resisting factor against gravity?
    a. weight
    b. cohesion
    c. friction
    d. strength of the slope material

20. Which of the following are most likely to act as triggers for a rapid mass wasting event?
    a. high temperatures and wind
    b. drought and wildfire
    c. friction and cohesion
    d. vibrations and increased water

21. The angle of repose for dry sand is
    a. 10 degrees.
    b. 25 degrees.
    c. 45 degrees.
    d. 65 degrees.

22. Of all the mass wasting processes, the one that is the slowest is
    a. slump.
    b. creep.
    c. debris flow.
    d. earthflow.

23. Freeze/thaw and wet/dry cycles can cause
    a. a creep.
    b. a slump.
    c. a lahar.
    d. earthflow.

24. The resistance that occurs when one object slides over another is known as
    a. strength.
    b. cohesion.
    c. inertia.
    d. friction.

25. The tendency for like particles to stick together is called
    a. strength.
    b. cohesion.
    c. inertia.
    d. friction.

26. A material's resistance to stress is called its
    a. strength.
    b. cohesion.
    c. inertia.
    d. friction.

Short Answer

1. The mass wasting process responsible for the movement of material in permafrost regions (areas in which water in the upper portion of the ground remains frozen for all or most of the time) on slopes of only a few degrees is called _____.

2. What are the differences between slides, falls, and flows?

3. How does water influence mass wasting?

4. Why is the topic of vegetation complex with regard to slope stability?

5. Explain how a trigger can initiate a mass wasting event.

6. Why is mass wasting called the "unsung hero" of geology?

STREAMS AND GROUNDWATER

Lesson 11

AT A GLANCE

Purpose

Learning Objectives

Materials Needed

Overview

Purpose

The activities in this lesson will lay the foundation for understanding stream and groundwater flow, and a river's impact on the landscape.

Learning Objectives

After completing this laboratory lesson, you will be able to:

- Calculate stream gradient and discharge.
- Describe how stream erosion and deposition have altered the landscape, and describe the depositional environments of various types of sediment.
- Describe the factors affecting the flow of groundwater.

Materials Needed

- ❏ Pencil
- ❏ Eraser
- ❏ Calculator
- ❏ Piece of string
- ❏ Ruler (in lab kit)
- ❏ USGS topographic map of Yosemite Valley (in lab kit)

Overview

The changes that occur as a result of plate tectonics are generally slow, visible only where there is a volcanic eruption or an earthquake that shifts the rock on either side of a fault. Far more rapid and visible over time are the changes caused by running water—the agent that sculpts canyons and river valleys, carrying to the ocean materials moved downward by mass wasting. Less visible and far slower than streams and rivers is the movement of groundwater beneath Earth's surface, which also moves downslope, carrying with it dissolved ions released during chemical weathering. Along the way, water nourishes life, forming environments where life's diversity can flourish.

Hydrologic Cycle

The water on Earth's surface is part of an Earth system called the hydrosphere. About 97 percent of the water in the hydrosphere is saltwater contained within Earth's oceans, which cover about 70 percent of Earth's surface. Oceans will be discussed in Lesson 12.

The remaining three percent of the water in the hydrosphere is fresh water. **Figure 11.1** shows the distribution of freshwater. What is astonishing to most individuals is the nearly negligible amount of water represented by all the rivers and streams of the world. Even the atmosphere contains more freshwater than all of Earth's rivers and streams, a fact that clearly points out the extreme efficiency of streams as the major agent of erosion. Equally surprising is the fact that most of Earth's freshwater is contained within the 12.9 million square kilometers (5 million square miles) of glacial ice that exists on Earth's surface (to be discussed in Lesson 13). Only 0.002 percent is used by the biosphere, which includes all living organisms on Earth.

Distribution of Earth's Freshwater		
Reservoir	**Volume** (10⁶ km³)	**Percent of Total**
Glaciers	29	74.05
Groundwater	9.5	24.58
Lakes	0.125	0.323
Atmosphere	0.013	0.034
Rivers	0.0017	0.004
Biosphere	0.0006	0.002
Total	38.64	99.00

Figure 11.1: Distribution of Earth's Freshwater. Illustration by Don Vierstra.

Freshwater moves about the earth, changing from one state to another in a continuous loop called the hydrologic cycle. The simplest way to describe its complexity is that water moves from the oceans to the continents and back again. During this journey, the water is sometimes in a gaseous state as water vapor, sometimes in a liquid state as freshwater or saltwater, and sometimes in a solid state as snow or ice.

Figure 11.2 illustrates the hydrologic cycle, which is usually described starting with the ocean. Solar energy evaporates ocean water so that it enters the atmosphere as water vapor. The winds transport the water vapor to a region of the atmosphere where the air is cooler, causing the water vapor to condense into rain or snow. Some of the rain or snow falls back into the ocean, completing the cycle, but a generous amount of precipitation falls on land. There, rainwater or snowmelt either evaporates, soaks into the soil, or runs off the surface, forming streams that flow down to the ocean. Some of the water that soaks into the soil is absorbed by plants, which release the water once again into the atmosphere in a process called transpiration. This water vapor may once again condense to fall as rain or snow. Other water soaks into the ground to become part of the groundwater system, where it may feed streams or lakes on its own journey toward the ocean. When the water from streams and groundwater reaches the ocean, it has completed the hydrologic cycle.

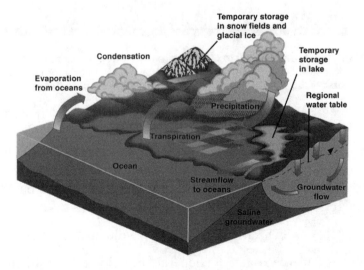

Figure 11.2 The Hydrologic Cycle. From *Planet Earth* by John J. Renton. Copyright © 2002 by John J. Renton. Reprinted by permission of Kendall Hunt Publishing Company.

Streams and Rivers

Approximately 25 percent of precipitation falling on the land becomes surface water, or runoff, and flows to the sea in river and stream channels. Along its journey, a stream changes character, a change geologists refer to as the age of a river. A stream begins as youthful, over distance it becomes mature, and as it approaches its final destination—a lake or the ocean—it becomes an old river. Each stage has different characteristics based upon a number of factors, including gradient (the steepness of the slope), the speed and manner in which it flows, and the type of deposition that occurs along the way. All of these factors are related to what geologists refer to as a stream's energy.

Hydrologists, geologists who study the systems involved in the movement, distribution, and quality of water, express the energy of a stream in terms of its discharge, the amount of water moving past a specific point in a stream over a specific time. Discharge is expressed as cubic meters per second or cubic feet per second. Two parameters needed to calculate discharge are volume and velocity. Both parameters are more complicated than they may seem at first glance because they vary so much, based on conditions.

The volume of a stream is actually expressed in terms of the volume of a cross section of the stream (**Figure 11.3**), not the entire stream. The cross section is either one meter (m) or one foot (ft) in length. The other two measurements in calculating volume are the depth of the water (called water height) and the width of the stream, so the formula to calculate volume is as follows.

Volume = height × width × length (always 1 m or 1 ft)

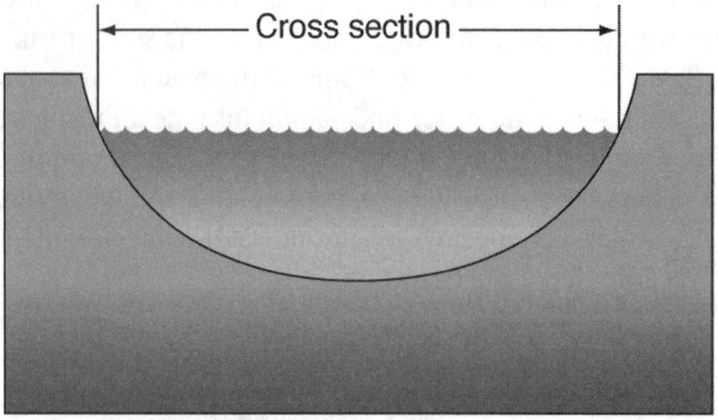

Figure 11.3 A Stream Cross Section. The volume of a stream is actually the volume of a one-foot or one-meter segment of the stream. It is one foot or one meter times the area of its cross section, which is the stream width times the water height. Illustration © Kendall Hunt Publishing Company.

Discharge is a measurement of how many of these cross sections would pass a specific point within a specified time, a value that changes with the stream's velocity.

Velocity is the speed at which the cross section moves downstream, and is measured by placing an instrument in the water at the same site as the cross section the hydrologist has used to calculate volume. The results are expressed in terms of cubic meters per second (m^3/s) or cubic feet per second (cfs). Velocity times volume is the stream's discharge.

Discharge = velocity × volume

Like volume, velocity is widely variable. In general, the velocity of a stream generally increases as it flows downstream due to the combined influence of three factors: (1) the gradient of the stream channel, discussed below, (2) the volume of water, which increases as tributaries join the main stream, and (3) the roughness of the stream channel. All three of these factors continually vary as the stream makes its journey toward its mouth.

Gradient is a term that refers to the slope of a stream section; in other words, the drop in elevation over a specified horizontal distance. **Figure 11.4** illustrates how gradient is calculated by dividing the difference in elevation between two points by the horizontal distance between the two points. The result is expressed either in meters per kilometer (m/km) or feet per mile (ft/mi).

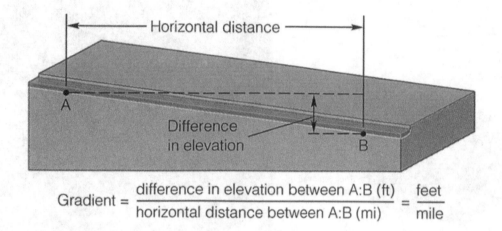

$$\text{Gradient} = \frac{\text{difference in elevation between A:B (ft)}}{\text{horizontal distance between A:B (mi)}} = \frac{\text{feet}}{\text{mile}}$$

Figure 11.4 Stream Gradient. One of the important stream parameters, the gradient, is the number of feet (or meters) the channel drops over a horizontal distance of one mile (or kilometer). Illustration by John J. Renton.

Higher gradients are generally found near a stream's source. In many cases, that is in the mountains. Intuitively, one would think that the high gradient of a rushing mountain stream or rapids would mean that the stream's velocity would be highest there, but this is not the case. The individual water particles may move more rapidly with the steep gradient, but they are also moving in a chaotic fashion referred to as a turbulent flow. Turbulence causes a particle of water to move along a topsy-turvy path, bouncing back upstream whenever it hits an obstacle or other water particles. This chaos slows down the progress of the water particle so that it is actually moving downstream at a slower velocity than it might appear.

Velocity is also slowed by the relative roughness of the channel bed. As a stream flows through bedrock, its particles abrade, or knock loose, rock particles from its channel. The abrasion produced during turbulence creates a rough channel bed that further slows the progress of any individual water particle. Velocity increases, however, with the volume of water. A stream swollen with meltwater in the spring moves more rapidly than one with a trickle of water, and, as noted earlier, as tributaries join the stream farther downstream, the stream's velocity increases with the added volume.

Stream Erosion and Deposition

All streams are powerful agents of erosion, downcutting into solid rock to produce a progressively deeper channel (see **Figure 11.5**). As the channel deepens, mass wasting causes the banks to collapse and the stream carries away the sediment that has moved downslope. This process repeats as the stream continues to downcut into the bedrock. Over time, the process produces a V-shaped stream valley, often referred to as a canyon.

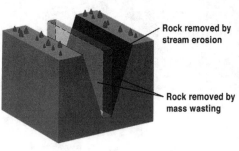

Rock removed by
stream erosion

Rock removed by
mass wasting

Figure 11.5 The V-shaped Stream Valley. Illustration by John J. Renton (left); Photo:
Shutterstock 70713988, credit Martin Lehmann (right).

Streams carry large amounts of sediment, most of which will be deposited somewhere along
their length. A stream's transportation of sediment is described in three ways: competence,
capacity, and load.

Competence is the largest size particle a stream is capable of transporting at a particular time.
Competence varies directly with velocity. If velocity decreases, competence decreases as well,
and deposition begins. If water velocity increases, so does competence and erosion occurs.
Capacity is the total amount of material the stream *can carry* at a given point, and load is the
amount of material the stream actually *is carrying* at any given time. There are three types of
load: suspended load, dissolved load, and bed load.

- Suspended load includes the particles the stream has picked up as it eroded the bedrock
  or washed away earth materials moved downslope by mass wasting, and can include a
  variety of particle sizes, depending upon the energy of the stream. These particles are
  suspended in the water in the same way chocolate powder is suspended in chocolate
  milk. Just as suspended chocolate powder settles out of milk to coat the bottom of the
  glass if the milk not frequently stirred, suspended sediments will settle to the stream
  bottom if turbulence diminishes and the stream becomes calmer.
- Dissolved load includes the substances that are dissolved in the stream's water; these
  may precipitate out farther downstream or after they reach the ocean.

- Bed load refers to rock particles that too large for the stream to carry as suspended load. Even though the particles in bed load are too large to be picked up, these particles do move downstream. They roll, they slide, or they bounce downstream; the bouncing is called saltation.

The higher the energy of a stream, the greater the range of particle sizes it can pick up and transport. When it is high-energy, a stream can carry particles as large as gravels in its suspended load. Deposition begins when the stream loses energy. The loss of energy could occur as a result of a decrease in gradient, which could be permanent or only for a short distance, or because the volume of water decreased. Streams often have high energy when they contain large volumes of water from melting snow or heavy rainfall and significantly lower energy at other times.

Streams deposit the sediment they carry in reverse order of size, depositing the gravels first as they begin to lose energy, then the sand-sized particles as energy continues to decrease. Gravels are mostly deposited in the streambeds of higher gradient streams after water volume has decreased enough for the energy to drop. Sand-sized particles are also deposited in these locations as well as farther downstream wherever energy drops enough for them to settle. Silts and clays are deposited last, usually toward the mouth of the river.

Erosion and deposition continues along the length of the stream. As tributaries join the stream, the added volume of water and resulting increased velocity also widens the stream channel. As the gradient becomes gentler, the stream becomes less turbulent and begins to meander, or flow in a sinuous path. **Figure 11.6** illustrates the meandering river. Meandering occurs because the stream flows at a higher velocity on the outside of a bend than it does on the inside of a bend or in the reaches, the relatively straight segments between bends. The higher velocity water on the outside of the bend erodes into the landscape, creating a cut bank, while the velocity of the water on the inside of the bend is slow enough to deposit some of the sand-sized particles in the suspended load, creating a point bar. Point bars are often favorite campsites or fishing spots because of the sandy beaches they offer, although point bars also contain gravel-sized particles transported there as bed load.

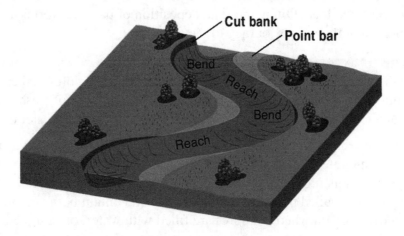

Figure 11.6 Meandering River. A meandering river forms cut banks on the outside of bends and point bars on the inside of bends. Illustration by John J. Renton.

Cut banks and point bars develop simultaneously. As the cut bank creates additional channel area on the outside of the stream, the point bar fills it in on the inside of the stream. The result is that the channel migrates in the direction of the cut bank. Meanwhile, the same process is occurring along other bends, including those turning in the opposite direction.

Over time, the meandering and migrating river becomes even more sinuous, winding back and forth over a wider area, cutting into the valley sides. Mass wasting then occurs, moving down material that is then carried away by the river. The valley widens as the stream meanders, and the slopes along the valley walls become gentler as mass wasting eliminates sharp ridgelines.

River valleys are subject to flooding during periods of high water, caused by excessive rainfall or snowmelt upstream. When water floods over the riverbank, it spreads out and forms a broad but relatively thin layer of water that moves much more slowly than the water in the channel. The lower velocity of this layer decreases its competence, so sediment, usually sand, settles out along the stream channel. Over time, this deposition of the sand-sized particles may build a levee. Meanwhile, the finer grained silts and clay-sized particles are carried out onto the floodplain and deposited there as the water velocity continues to slow. If you picture a flooded region, generally the water is moving very slowly (except near a levee breach).

Repeated floods fill the valley near the river with silt- and clay-sized sediments, creating a floodplain, a term that refers to the area covered by the largest floods. The deposition also creates fertile soil, and many farmlands are located in current and former floodplains. Eventually, the sediments deposited during floods may once again be picked up by the water and carried downstream to the stream's mouth.

As the stream approaches the end of its journey, the stream gradient and turbulence are very low, and the river valley is wide enough so that it is many times wider than the stream channel. Here, the landscape takes on the appearance of a broad, flat surface underlain with silts and clays deposited by floods. Meanwhile, the levees built of sand-sized particles are likely to be much higher.

As the waters of the stream approach the mouth, one of the stream's major tasks, the transport of sediment, has been fulfilled. At this point, most of the sands have been deposited and, except during floods, the stream's suspended load is almost completely silt and clay-sized particles; these settle out as the stream enters a larger body of water whose currents are not strong enough to carry the load. Often, the final deposition of particles forms a delta that makes the completion of the waters journey to the sea.

Groundwater

Groundwater is the water that is contained underground within the pores and crevices of soil, sediment, and solid rock. As noted above in **Figure 11.1**, the volume of freshwater within the ground is estimated to be as much as forty times more than all the freshwater on the surface.

Freshwater enters the groundwater system when rainwater or melting snow infiltrates, or soaks into, the ground and migrates downward. The water passes first through the top layer of soil, called the zone of soil moisture (**Figure 11.7**), then through the zone of aeration where the spaces between particles, called pores, in sediment and soil contain both air and water. Eventually it reaches a region where all pores are filled with water, or saturated. This is the zone of saturation. The boundary between the zone of saturation and the zone of aeration is called the water table.

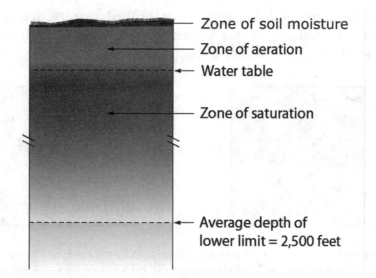

Figure 11.7 Water Table. The water table is the boundary between the zone of saturation and the zone of aeration. Illustration © Kendall Hunt Publishing Company.

The depth to the water table (also called its height) varies from season to season and with the presence of drought or high precipitation. The water entering the groundwater system from rain, snowmelt, or from streams is called recharge. Water leaving the groundwater system via springs, lakes, streams, or pumping is called discharge.

When recharge exceeds discharge, as it does during wet periods (meaning years of heavy rainfall) or after a snowmelt, the water table rises. In this case, the water table may intersect with the surface, creating springs, swamps, and spring-fed lakes where it does. When discharge exceeds recharge, as it does during a drought, the water table drops. Springs, lakes, and swamps may dry up and disappear, and water from streams above the water table will lose water as it percolates down to the groundwater system. Either way, the water table is not at a uniform depth but follows the surface topography, being slightly elevated under hills and depressed under valleys across a region (**Figure 11.8**).

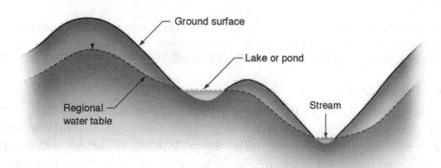

Figure 11.8 Regional Water Table. Illustration © Kendall Hunt Publishing Company.

Groundwater is contained in soil, sediment, or rock. Rather than being complete solid, most rocks contain some open spaces that can hold water. In some cases, the spaces are the result of the pores that exist between the individual mineral grains; in other cases, fractures create spaces. In the case of the carbonate rocks, the fractures may be enlarged because carbonates tend to dissolve in acidic water, and rainfall is slightly acidic. The dissolution creates solution channels. The percent of rock volume that is space is called porosity. Porosity (**Figure 11.9**) is important because it determines how much water the rock will hold. Different rock types exhibit different levels of porosity, ranging from 0 to 50 percent.

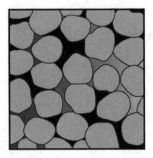 

Figure 11.9 Porosity. The porosity of a rock is determined by the combined space between mineral grains (left) and within fractures (right). Illustration © Kendall Hunt Publishing Company.

Even if a rock is porous, water may not flow through it unless the spaces are connected. The ability of a rock to transmit fluids is called permeability (**Figure 11.10**). Rocks with low porosity almost always have low permeability as well. Rocks with high porosity have high permeability as long as the spaces are connected. Vesicular basalt is an example of a rock that is porous but not very permeable, as the vesicles are not usually connected to each other. Shales are also porous but have low permeability.

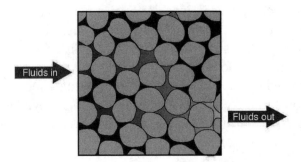

Figure 11.10 Permeability. Permeability refers to the extent to which water can flow through rock. High-permeability rocks allow for efficient water flow. When saturated, these rocks are known as aquifers. Low-permeability rocks impede the flow of water and are known as aquitards. Illustration © Kendall Hunt Publishing Company.

Aquifers and Aquitards

Different geologic materials have different porosities and permeabilities and so can hold and transmit different volumes of groundwater. For instance, loose, unconsolidated sediment, such as sand, has a higher porosity and permeability than solidified sandstone, because the compaction and cementation that accompanies lithification reduces the amount of pore space in the sandstone. Consequently, unconsolidated sand would hold and transmit more water than sandstone.

An aquifer is a body of rock that is porous and is also permeable enough to conduct economically significant quantities of water to wells and springs. On the other hand, an aquitard either has low porosity and permeability, or it is porous but not permeable. Igneous rocks, metamorphic rocks, and shale are typical aquitards. Sandstone, coarse-grained sediment, and limestone with solution channels are examples of good aquifers.

Unconfined Aquifer

Like water in a stream, groundwater flows. The movement of groundwater is driven by gravity and directed by the presence of aquifers and aquitards. That, in turn, affects the ability of wells to access the water.

An unconfined aquifer is one through which there is a direct vertical connection from the water table to the atmosphere through pore spaces; in other words, there is no aquitard between the aquifer and the surface. Unconfined aquifers are so named because once the water enters the aquifer, it is *not confined* to it aquifer but may enter another aquifer above or below it by moving through fractures or pores within the rock section. Once in an aquifer, the water will flow in the same general direction as the overall surface drainage down a gradient or slope.

A characteristic of wells drilled into unconfined aquifers is that the water in the well will rise only to the level of the water table and a pump is necessary to lift the water to the surface. Pumping water faster than it can be replaced will create a cone of depression in the water table, thus causing a lowering of the water table in the vicinity of the well (**Figure 11.11**). Continued overproduction causes the cone of depression to expand until a point of equilibrium is reached in which the pump only yields as much water as is being replaced.

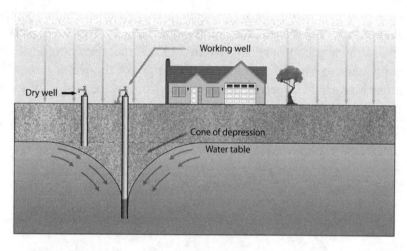

Figure 11.11 Cone of Depression in an Unconfined Aquifer. Illustration by Don Vierstra.

Water withdrawn from a well in an unconfined aquifer creates a cone of depression, a cone-shaped drop in the water table around the well.

Confined Aquifer

An aquifer sandwiched between two confining beds is a confined aquifer (see **Figure 11.12**), because the confining beds prevent free vertical movement of the groundwater. Because of restricted vertical movement, water pressure in a confined aquifer can build to high levels, especially if the groundwater is being replenished at higher elevations in the recharge area and the confined aquifer is almost fully saturated with groundwater. If the upper aquitard should fracture, or if a well is drilled through it, the high pressure causes the water to rise to a height well above the confined aquifer. It may even reach the surface with enough pressure to flow freely. Such a condition is known as an artesian system; the break in the aquitard creates a free-flowing artesian spring and a well drilled through the aquitard is a free-flowing artesian well. If pressure is not enough to force the water to the surface, a pump may bring it up the rest of the way. In this case, the well is considered non-free-flowing.

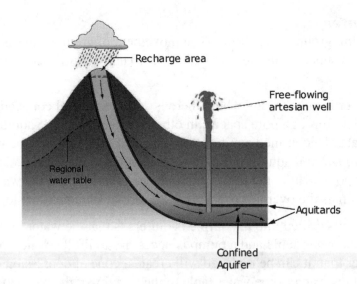

Figure 11.12 Artesian System. Illustration © Kendall Hunt Publishing Company.

Groundwater Flow in Unconfined Aquifers

Groundwater, like stream water, is on a journey that will ultimately terminate in the ocean (although it may join a stream along the way). Generally, groundwater flows downhill, just as a stream does, away from the recharge area and toward the discharge area. This flow is illustrated with a groundwater contour map (**Figure 11.13**) that uses contour lines similar to those in topographic maps to show the change in the height of the water table. The groundwater contour is overlaid over a surface map and shows the height of the water table beneath the surface. Each contour line in **Figure 11.13** represents a change of one foot in the height of the water table. Groundwater flows perpendicular to the contour lines in the direction of the arrows.

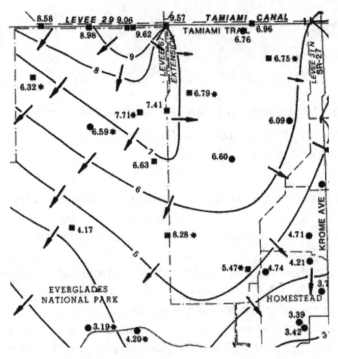

Figure 11.13 Groundwater Contour Map. In this groundwater contour map of a small portion of Florida's Dade County, each contour line in represents a change of one foot in the height of the water table. Groundwater flows perpendicular to the contour lines in the direction of the arrows. Map: Courtesy of USGS.

The speed at which groundwater flows is typically very slow compared to the stream velocities. (The exception is in areas where the aquifer is limestone with many solution channels, underground rivers, and caverns.)

The actual rate of groundwater flow depends on the porosity and permeability of the rock and the slope of the water table. In extremely porous and permeable sands and gravels, the water may move at rates of 12 to 15 centimeters (5 to 6 inches) or more per day. At the other extreme, such as water in an aquitard like fine-grained shale, water may flow at rates of a few centimeters per year, especially if the slope of the aquifer is low. On the average, groundwater moves through most aquifers at a rate of about an inch or so per day. One of the results of such a slow movement of groundwater is that the water from a well or spring may have been in the ground for hundreds or even thousands of years. On a more practical level, the rate of flow determines the rate at which water can be pumped from a well without depleting the aquifer. This information is used to calculate safe production from wells.

The rate of movement also determines the spread of groundwater contamination. While the purity of groundwater is normally considered to be equal to that of rainwater, contamination can be introduced from a wide variety of sources. Most water pollution stems from human activities. In rural areas, groundwater provides water for domestic and farm use, so major sources of contamination may be human sewage, livestock waste, pesticides, and fertilizers. In urban areas, the contamination is likely to come from industrial processes that have been improperly disposed. Some of these substances are harmful to human health.

Fortunately, sediments and rocks contained in the zones of aeration and saturation have an impressive ability to remove contaminants from even the most polluted waters. Most sand-sized materials can purify groundwater of organic contaminants within a few hundred meters of the point of entry, although many industrial contaminants are not so easily removed. Nevertheless, there is a limit to what the natural purification system can accomplish. When locating wells to be used as sources of water for human consumption, it is important to consider both the geology below ground and any surrounding potential sources of pollution.

Lab Exercises

Lab Exercise #1: *Yosemite Valley*

In this laboratory exercise you will apply what you have learned to answer some questions about the Merced River, which flows through Yosemite Valley. Follow the instructions below and record your results in the space provided or on a separate piece of paper. Make sure to save your results. You will record your answers to the questions in the quiz at the end of this lesson.

Instructions and Observations

Step 1: Lay out the USGS topographic map of Yosemite Valley found in your lab kit.

Step 2: Measure the distance along the Merced River from the Leidig Meadow footbridge to El Capitan Bridge. The easiest way to do this is to use a string. Position the string so that one end is at the Leidig Meadow footbridge and the other is at El Capitan Bridge, making sure that the string follows all of the bends in the river. Mark the string at the El Capitan Bridge. Then compare the length noted on the string to the scale on the topographic map to determine the distance covered.

If you would prefer to use a ruler or straight edge, the best approach is to measure the length of one-tenth mile using the mileage scale on the map. Use that to make tick marks along the river between the two bridges. When you have marked off the entire distance, multiply the number of tick marks from one bridge to another and multiply by 10 to get the result in miles. (Make sure not to count the starting mark!)

1. What is the distance in miles along the Merced River from the Leidig Meadow footbridge to the El Capitan Bridge?

Step 3: Find the approximate elevation of the Leidig Meadow footbridge and the El Capitan Bridge using the benchmarks along the road nearest to each bridge.

2. What is the difference in elevation between the Leidig Meadow footbridge and El Capitan Bridge?

Step 4: Use the values in Steps 2 and 3 to calculate the stream gradient of the Merced River between the Leidig Meadow footbridge and El Capitan Bridge.

3. What is the gradient?

Step 5: River height and width varies from week to week and often from day to day. On one particular day, the width of the Merced River at the gauge station at Pohono Bridge was 22 feet and the river's height at that location was 3.6 feet. Calculate the volume of a cross section of the Merced River at the Pohono Bridge on that day.

4. What was the volume of a cross section at this location on that day?

Step 6: Assuming that the river's velocity at Pohono Bridge was 6 ft/sec on that day, calculate the river's discharge. The formula appears in this lesson's Overview.

5. What was the river's discharge?

Step 7: Rising waters along the Merced River have an impact on the Yosemite Valley in years with high rainfall/snowfall and the resulting spring snowmelt. Examine the Yosemite Valley Map and answer the following questions:

6. Which of the following is most likely to be closed as a result of flooding?
   a. the one-way road to the south of Yosemite Lodge
   b. the one-way road to the north of Yosemite Lodge
   c. the tunnel on Big Oak Flat Road
   d. Wawona Road

7. If flooding were imminent, which of the following locations appears to be the safest?
   a. Yosemite Village Visitor's Center
   b. Riverside Camp
   c. Pines Camp
   d. the Chapel

Step 8: There are several gauging stations in Yosemite Valley placed there to measure the flow of the rivers. The normal river height of the flow at the Pohono Bridge Gauge is 3.5 feet. The river is considered to be at flood stage when it is 6.5 feet above this level, or 10 feet. During a record flood, the water level reached 23.45 feet. Happy Isles Gauging Station is at an elevation of 4017 feet located where the Little Merced River enters the Valley.

Look at the topographic map and list a minimum of two potential problems you can deduce would be caused by the record flooding.

8. What are two potential problems that would be caused by a record flood?

9. During flood conditions, would you expect the river height recorded at Happy Isles to be greater, equal to, or less than the height recorded at Pohono Bridge? Why?

Lab Exercise #2: *Stream Deposition*

In this laboratory exercise, you will apply what you've learned about stream deposition to the clasts you examined in **Lesson 7, Lab Exercises #1 and #2**. Each of the samples could have been deposited in one or more typical stream environments. Your task is to use what you've learned about deposition and particle sizes in this lesson to determine the environments in which each sample is likely to have been deposited.

Follow the instructions below and record your results in the space provided or on a separate piece of paper. Make sure to save your results. You will record your answers in the quiz at the end of this lesson.

Instructions

Step 1: Retrieve your answers from **Figures 7.7 and 7.8**, then choose the most likely depositional environments from the list below. Some sediments can be deposited in a wide variety of settings but dominate some settings more than others, so answer these questions in terms of where they are *most likely* to be found, not where it is *possible* for them to be found. If you have difficulty answering the questions below based on your answers from Lesson 7, feel free to reexamine the sediment and sedimentary rock samples.

Depositional environments:

mountain streambed   floodplain
point bar   delta
levee

10. In which of the following stream environments were the sediments in Bag A *most likely* to have been deposited?

    a. mountain streambed

    b. point bar

    c. floodplain

    d. delta

11. In which of the following stream environments were the sediments in Bag B *most likely* to have been deposited?

    a. mountain streambed, floodplain

    b. mountain streambed, point bar, levee

    c. levee, delta

    d. mountain streambed, delta

12. In which of the following stream environments were the sediments in Bag C *most likely* to have been deposited?

    a. mountain streambed

    b. floodplain

    c. delta

    d. levee

13. In which of the following stream environments were the sediments in Bag D *most likely* to have been deposited?

    a. mountain streambed

    b. point bar

    c. levee

    d. delta, floodplain

14. In which of the following stream environments were the sediments in Bag E *most likely* to have been deposited?

    a. delta, levee

    b. floodplain, point bar

    c. mountain streambed, point bar

    d. levee, floodplain

15. In which of the following stream environments were the sediments composing sedimentary rock specimen #11 *most likely* to have been deposited?

    a. delta

    b. floodplain, point bar

    c. levee, floodplain

    d. mountain streambed, point bar, levee

16. In which of the following stream environments were the sediments composing sedimentary rock specimen #13 *most likely* to have been deposited?

    a. mountain streambed

    b. floodplain

    c. levee

    d. delta

17. In which of the following stream environments were the sediments composing sedimentary rock specimen #17 *most likely* to have been deposited?

    a. point bar, levee

    b. delta, floodplain

    c. mountain streambed, point bar

    d. levee, floodplain

Lab Exercise #3: *Groundwater Flow*

In this laboratory exercise, you will apply what you've learned about groundwater flow to answer questions about possible water contamination at a strip mall. Follow the instructions below and record your results in the space provided or on a separate piece of paper. Make sure to save your results. You will record your answers in the quiz at the end of this lesson.

Instructions

Step 1:   **Figure 11.14** shows a typical strip mall in a typical city, except that it has been identified as a potential source of groundwater contamination. Only preliminary assessments have taken place to determine the height of the water table (in feet relative to mean sea level) and the location of two Underground Storage Tanks (USTs) owned by the gas station located at the site. Other businesses at the site include a pizza parlor, a convenience store, and a liquor store.

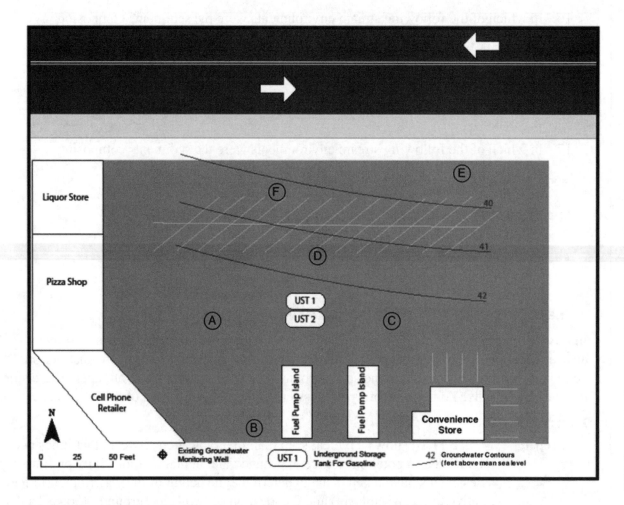

Figure 11.14 Possible Groundwater Contamination. Illustration by Roberto Falero.

As an environmental consultant you've been hired to write a work plan to assess possible contamination at the site. In your report, you answer the following questions.

18. If there is groundwater contamination at the site, what would be the most likely source?

    a. the cell phone retailer

    b. the fuel pump islands

    c. the underground storage tanks (USTs)

    d. the liquor store

19. Provided the source of any contamination is this strip mall in **Figure 11.14**, what is the contamination most likely to be?

    a. liquor

    b. toxins from cell phone batteries

    c. gasoline

    d. food waste from the Pizza Shop

The flow of groundwater is perpendicular to the contour lines. Which direction is groundwater flowing beneath the strip mall? Use your compass and the methods discussed in Lesson 1 to find a compass bearing.

20. In which compass direction is the groundwater flowing?

    a. N10°W

    b. N10°E

    c. N80°E

    d. N80°W

21. Which two locations—A, B, C, or D—would be the best places to install groundwater monitoring wells?

    a. A and B

    b. B and F

    c. C and E

    d. B and D

22. Why would the locations you chose in **Figure 11.14** be the best places to install groundwater monitoring wells?

Online Activities

Per your instructor's directions, go to the online supplement for this lab and complete the activities assigned. Viewing the online videos will help you to complete the quiz.

Quiz

Multiple Choice

1. All of the following stream parameters generally increase downstream EXCEPT for

    a. gradient.

    b. discharge.

    c. load.

    d. capacity.

2. Record your answer to **Lab Exercise #1, Question 1**. What is the distance in miles along the Merced River from the Leidig Meadow footbridge to the El Capitan Bridge?

    a. 0.5 miles

    b. 2.7 miles

    c. 3.7 miles

    d. 5.9 miles

3. Record your answer to **Lab Exercise #1, Question 2**. What is the difference in elevation between the Leidig Meadow footbridge and El Capitan Bridge?

    a. 5 feet

    b. 10 feet

    c. 20 feet

    d. 40 feet

4. Record your answer to **Lab Exercise #1, Question 3**. What is the stream gradient of the Merced River between the Leidig Meadow footbridge and El Capitan Bridge?

    a. 3.70 feet per mile

    b. 6.70 feet per mile

    c. 12.70 feet per mile

    d. 0.27 feet per mile

5. Record your answer to **Lab Exercise #1, Question 4**. What was the volume of a cross section at Pohono Bridge on the day its width was 22 feet and its water height was 3.6 feet?

    a. 60 square feet

    b. 79.2 square feet

    c. 79.2 cubic feet

    d. 105 cubic feet

6. Record your answer to **Lab Exercise #1, Question 5**. Assuming that the river's velocity at the gauge station was 6 ft/sec on that day, what was the discharge of the Merced River at Pohono Bridge?

    a. 300 cfs

    b. 475 cfs

    c. 535 cfs

    d. 790 cfs

7. Record your answer to **Lab Exercise #1, Question 6**. If the Merced River reached flood stage, which of the following is most likely to be closed as a result of flooding?

    a. the one-way road to the south of Yosemite Lodge

    b. the one-way road to the north of Yosemite Lodge

    c. the tunnel on Big Oak Flat Road

    d. Wawona Road

8. Record your answer to **Lab Exercise #1, Question 8**. If the Merced River reached flood stage, which of the locations appears to be the safest?

    a. Yosemite Village Visitor's Center

    b. Riverside Camp

    c. Pines Camp

    d. the Chapel

Questions 9 through 16 are based on **Lab Exercise #2:** *Stream Deposition.*

9. Record your answer to **Lab Exercise #2, Question 10**. In which of the following stream environments were the sediments in Bag A *most likely* to have been deposited?

    a. mountain streambed

    b. point bar

    c. floodplain

    d. delta

10. Record your answer to **Lab Exercise #2, Question 11**. In which of the following stream environments were the sediments in Bag B *most likely* to have been deposited?

    a. mountain streambed, floodplain

    b. mountain streambed, point bar, levee

    c. levee, delta

    d. mountain streambed, delta

11. Record your answer to **Lab Exercise #2, Question 12**. In which of the following stream environments were the sediments in Bag C *most likely* to have been deposited?

    a. mountain streambed

    b. floodplain

    c. delta

    d. levee

12. Record your answer to **Lab Exercise #2, Question 13**. In which of the following stream environments were the sediments in Bag D *most likely* to have been deposited?

    a. mountain streambed

    b. point bar

    c. levee

    d. delta, floodplain

13. Record your answer to **Lab Exercise #2, Question 14**. In which of the following stream environments were the sediments in Bag E *most likely* to have been deposited?

    a. delta, levee

    b. floodplain, point bar

    c. mountain streambed, point bar

    d. levee, floodplain

14. Record your answer to **Lab Exercise #2, Question 15.** In which of the following stream environments were the sediments composing sedimentary rock specimen #11 *most likely* to have been deposited?

    a. delta
    b. floodplain, point bar
    c. levee, floodplain
    d. mountain streambed, point bar, levee

15. Record your answer to **Lab Exercise #2, Question 16.** In which of the following stream environments were the sediments composing sedimentary rock specimen #13 *most likely* to have been deposited?

    a. mountain streambed
    b. floodplain
    c. levee
    d. delta

16. Record your answer to **Lab Exercise #2, Question 17.** In which of the following stream environments were the sediments composing sedimentary rock specimen #17 *most likely* to have been deposited?

    a. point bar, levee
    b. delta, floodplain
    c. mountain streambed, point bar
    d. levee, floodplain

Questions 17 through 20 are based on **Lab Exercise #3:** *Groundwater Flow.*

17. Record your answer to **Lab Exercise #3, Question 18.** If there is groundwater contamination at the site, what would be the most likely source?

    a. the cell phone retailer
    b. the fuel pump islands
    c. the underground storage tanks (USTs)
    d. the liquor store

18. Record your answer to **Lab Exercise #3, Question 19.** Provided the source of any contamination is the strip mall in **Figure 11.14**, what is the contamination most likely to be?

    a. liquor
    b. toxins from cell phone batteries
    c. gasoline
    d. food waste from the Pizza Shop

19. Record your answer to **Lab Exercise #3, Question 20.** In which compass direction is the groundwater flowing?

    a. N10°W
    b. N10°E
    c. N80°E
    d. N80°W

20. Record your answer to **Lab Exercise #3, Question 21**. Which two locations—A, B, C, or D—would be the best places to install groundwater monitoring wells?

    a. A and B
    b. B and F
    c. C and E
    d. B and D

21. What percentage of precipitation falling on the land becomes surface water, or runoff, and flows to the sea in river and stream channels?

    a. 10%
    b. 15%
    c. 20%
    d. 25%

22. The volume of freshwater held beneath the earth's surface constitutes as much as _____ the freshwater contained in lakes, rivers, and streams.

    a. 10 times
    b. 20 times
    c. 40 times
    d. 80 times

23. An aquifer must be

    a. porous and impermeable.
    b. porous and permeable.
    c. impermeable and nonporous.
    d. just impermeable.

24. Of the following, the best aquifer would be

    a. a well-sorted sand.
    b. a well-cemented sandstone.
    c. a siltstone or shale.
    d. an igneous rock such as granite.

25. A confined aquifer is always confined between

    a. the water table and the earth's surface.
    b. two impermeable layers.
    c. permeable layers.
    d. a permeable layer and impermeable layer.

26. The water table rises when

    a. recharge and discharge are about equal.
    b. discharge exceeds recharge.
    c. recharge exceeds discharge.
    d. more than one pump is located in the recharge area.

27. What factors affect the rate of groundwater flow?

    a. the presence of industrial contamination
    b. higher than usual rainfall or snowmelt
    c. porosity and permeability of the aquitard and slope of the water table
    d. porosity and permeability of the aquifer and slope of the water table

28. Where do springs, swamps, and spring-fed lakes occur?

    a. where the water table intersects with the surface
    b. in the recharge area
    c. only in the lower elevations
    d. downslope of a stream

29. Streams deposit sediment

    a. as they lose energy.
    b. in order of size, smaller particles first.
    c. when the suspended load exceeds the bed load.
    d. by a & b.

30. Which of the following statements is true about a stream valley?

    a. It forms through a combination of downcutting and mass wasting.
    b. It forms only where the bedrock is relatively soft.
    c. It is typically flat and wide where gradients are higher.
    d. a & c

Short Answer

1. Record your answer to **Lab Exercise #1, Question 8**. If the Merced River reached record flood heights, what are two potential problems that might that cause in Yosemite Valley?

2. Record your answer to **Lab Exercise #1, Question 9**. During record flood conditions, would you expect the river height recorded at Happy Isles to be greater, equal to, or less than the height recorded at Pohono Bridge? Why?

3. Record your answer to **Lab Exercise #3, Question 22**. Why would the locations you chose in **Figure 11.14** be the best places to install groundwater monitoring wells?

4. Describe what causes a stream to meander, forming cut banks and point bars as it does.

5. What geologic conditions are necessary for artesian wells?

6. What happens to the cone of depression if pumping increases?

7. How can a confined aquifer become contaminated?

OCEANS AND COASTLINES

Lesson 12

Purpose

Many urban areas have developed in coastal regions, causing populations to grow exponentially in such proximity to our coastlines. Coastlines are dynamic features that are under constant change as a result of erosional and depositional processes. In this lesson, you will learn how about the types of processes that change the contours of our coastlines.

Learning Objectives

After completing this laboratory lesson, you will be able to:

- Describe how wind-driven waves and the tides form and behave.
- Explain the difference between an emergent and submergent coastline.
- Name at least three erosional and three depositional coastline features, and describe how each forms.
- Calculate the tidal range in a coastal zone community.

Materials Needed

- ❏ Pencil
- ❏ Eraser
- ❏ Calculator

Overview

The ocean covers 70 percent of Earth's surface. Its depths only began to be explored about 150 years ago, but its coastlines have been mapped for centuries. While the study of the ocean is vast enough to support an entire field of study, known as oceanography, this lesson narrows that focus to coastlines and the waves and currents that continually change their contours.

Water constantly moves throughout all of the ocean basins. Many of us have enjoyed watching the seemingly constant arrival of waves at the shoreline, the distance they travel inland varying with the tides. Waves erode rocky shorelines, carving into them distinctive erosional features. Longshore currents pick up sediments and carry them along the shoreline only to deposit the sediments elsewhere to form depositional features. These processes make coastlines and shorelines areas of constant change.

The terms *coastline* and *shoreline* are often used interchangeably; however, they do not refer to the same areas. A shoreline is the area of land between the highest possible water level, which occurs during storms, and the lowest water level, which occurs during the lowest tides. By contrast, a coastline is the total area of interaction between land and sea, including seaside homes, wetlands along an ocean shore, and delta areas built up by freshwater rivers that deposit sediment at their mouths.

Wind-Driven Waves

Waves are the major agents of change along the shorelines. In some cases, the change is erosional as the waves sculpt the rocks, constantly undercutting the coast, driving sea cliffs landward and removing the coastal sediments away from the shoreline. In other cases, the action of the waves is depositional as the currents along the shore deposit sediments to build a variety of features.

The waves that continuously arrive at shorelines are wind-driven waves. They form when wind moves across the surface of the water, transferring energy from the moving air mass to the water. The peaks of the waves are referred to as crests; the lowest part of the depression between each crest is called the trough. Each wave is characterized by a wavelength (the distance between successive crests or troughs) and a wave height, the total distance from the top of a crest to the bottom of the adjacent trough. Another measurement often used is amplitude, which is the distance from a midpoint to the crest or trough of the wave (**Figure 12.1**). The number of wavelengths that pass a fixed point over specific time period is called the wave period.

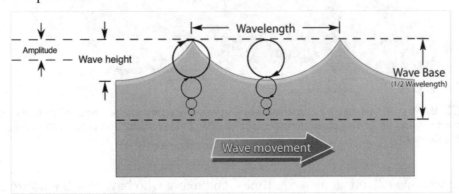

Figure 12.1 Wavelength. In water deeper than the wave base, a passing waveform results in no lateral motion of the water. Above that depth, the water is moved in circular paths that decrease in diameter as depth increases. Illustration by Don Vierstra.

As they form, the waves move in the direction of the wind that formed them until they lose enough energy to dissipate or until they reach the shoreline. It's important to note that what moves through the ocean is energy, not water particles. The water particles within a wave have a more limited motion, moving in a circle as the crest and trough of each wave passes through (**Figure 12.1**). Think of how a cork bobs in water moves as a wave passes through: up, forward, down, and back, in a circle. Water beneath the wave also moves in circles that diminish in diameter until a depth that is half of the wave's wavelength, known as the wave base. Water below the wave base does not move at all as the result of the wave passing through.

The height of waves depends on three factors: the velocity of the winds, the duration of the winds, and the fetch—the distance over which the winds blow. These are all direct relationships: the greater the wind's velocity, its duration, and its fetch, the greater the wave height. These three factors explain how an ocean location can be calm and glassy one day, and have high waves on another. Tropical storms and hurricanes have high-velocity winds that blow over large areas for a long time; consequently, such storms can generate very high and dangerous waves.

Once the waves are generated, they travel across the open ocean as successive wave crests, dissipating their energy as they do. Physical geologists are primarily interested in the portion of the original wave energy remaining that is available to erode and modify the coastline when the waves reach the shore.

As the wave approaches the shore, it will eventually reach a depth equivalent to its wave base; then the water particles moving in circular movements at the bottom of the wave base begin to touch bottom. At this point, the wave is said to "feel" bottom, as friction causes the wavelength to shorten and the crests to arrive in closer succession. At the same time, the wave height increases and the wave becomes asymmetric (see **Figure 12.2**). As the wave continues to change in height and shape, it becomes unstable, and it becomes a breaker as faster moving water at the top curls forward, then crashes down as surf. Whatever energy remains in the surf drives the water up onto shoreline, where the original energy is eventually consumed. Gravity takes over, returning the water to the sea as backwash. It is the energy delivered to the shoreline by surf that drives coastal processes such as coastal erosion and deposition.

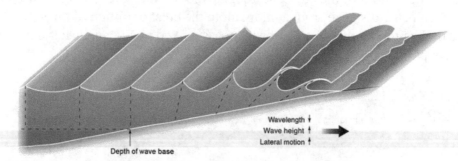

Figure 12.2 Wave Movement. When an arriving wave reaches a depth equivalent to its wave base, its wavelength shortens and its wave height increases until it is no longer stable and breaks. Illustration © Kendall Hunt Publishing Company.

Tides

Most people who have lived along a coastline—and many who haven't—are aware of the tides, a type of wave that continuously changes sea level along a shoreline. Tides are the result of gravitational forces exerted on Earth by the moon and, to a lesser extent, the sun. The forces form a bulge of ocean water on the side of Earth facing the moon (**Figure 12.3**). At the same time, complex interactions create a second bulge on the far side of the planet.

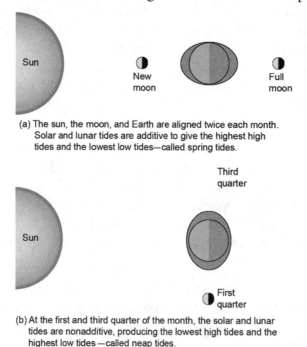

(a) The sun, the moon, and Earth are aligned twice each month. Solar and lunar tides are additive to give the highest high tides and the lowest low tides—called spring tides.

(b) At the first and third quarter of the month, the solar and lunar tides are nonadditive, producing the lowest high tides and the highest low tides—called neap tides.

Figure 12.3 Effect of the Sun and the Moon on Monthly Tidal Changes. The monthly tidal range is the result of the changing relationship between the sun, Earth, and the moon. When all three bodies are aligned along the same axis, as they are during the new and full phases of the moon, the tidal effects of the moon and sun are additive, resulting in the maximum tidal range, the so-called spring tide. When the moon is in either the first or third quarter phase, the tides are largely controlled by the moon and the tidal range will be at its minimum, the so-called neap tide. From *Planet Earth* by John J. Renton. Copyright © 2002 by John J. Renton. Reprinted by permission of Kendall Hunt Publishing Company.

The bulges are actually the crests of two waves, each with a wavelength of almost halfway around Earth's diameter. As Earth rotates beneath the crest of these long-wavelength waves, it appears to us that it is the wave that moves, circling Earth at a speed slightly slower than Earth's rotation, arriving at the same location 50 minutes later every day. The reason for the apparent delay is that over a 24-hour period, the moon has moved farther through its orbit around Earth and the tidal bulge is synchronized with the location of the moon.

When a given location moves beneath one of the bulges, sea level in that area slowly rises, and it is said to experience a flood tide. Sea level reaches its highest levels of the day when the location passes under the crests of the bulge, experiencing high tides. Likewise, twice a day, sea level will gradually decrease as the location experiences an ebb tide, the result of Earth's rotation moving the location out from under the bulge. Low tide occurs when sea level reaches its lowest point. Along most coasts, both high and low tides occur twice each day as Earth rotates through the two tidal bulges. Generally, one high tide will be higher than the other high tide, and one low tide will be lower than the other low tide.

While the moon's gravitational pull is the main force involved in creating the bulge, the gravitational attraction from the sun also plays a role, either adding to or partially canceling the gravitational pull of the moon. As noted in **Figure 12.3**, when the moon and sun line up in a row, as they do twice a month, their gravitational attractions combine to create a higher bulge, a condition known as a spring tide. During spring tides, the tidal range is at its maximum value; that is, highest tide of the day reaches its highest value of the month while the lowest tide of the day reaches its lowest value of the month. When the moon and sun are positioned at 90° angles from each other, which also happens twice a month, their gravitational attractions partially cancel each other out, creating a lower bulge, a condition known as neap tide. During neap tides, the tidal range is at a minimum—the highest tide of the day is at its lowest level of the month while lowest tide of the day is at its highest level of the month.

The tidal range experienced by a particular coastline is dependent on many factors, including the slope of the ocean bottom and the coastline's shape, including its inlets. In general, the tidal range is relatively small in areas such as the Hawaiian Islands or off the Florida Keys where the ocean bottom drops off steeply, and can be quite large where the ocean bottom slopes gently seaward. The greatest tidal range in North America is in the Bay of Fundy, Nova Scotia, Canada, where the tidal range can be as high as 18 meters (50 feet) (see **Figure 12.4**). The extreme tidal range within the Bay of Fundy is the result a number of factors, including the gently sloping bottom within the bay and the funneling of the waters through a restricted channel that connects the bay with the open sea.

Figure 12.4 Bay of Fundy Tidal Range. Because of its unique combination of a gentle offshore slope and confinement by the adjacent landmass, the Bay of Fundy, Nova Scotia, Canada, experiences the highest tidal range in North America. Photos: Courtesy of Nova Scotia Tourism.

Longshore Currents

A current is defined as the movement of a fluid; in the case of the ocean, the fluid is seawater. Prevailing winds, Earth's rotation, and density differences between saltwater and freshwater drive numerous currents throughout Earth's oceans. The current that has the most direct impact on the contours of a local shoreline, however, is the longshore current.

A longshore current is the result of waves breaking onshore. For the most part, waves approach the shoreline at an angle. When the waves break, they push water onshore at the angle of approach. The backwash, however, is perpendicular to the slope of the shoreline. The change in angles creates a zigzag pattern that moves water down the beach as a longshore current (**Figure 12.5**). The zigzag pattern also moves sediment down the beach. This sediment, transported by the longshore current, is called beach drift or longshore drift.

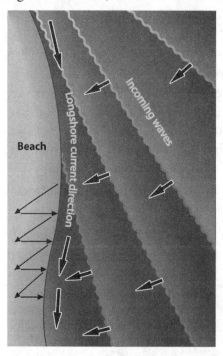

Figure 12.5 Longshore Current. Waves typically approach a beach from an angle, but the direction of the backwash is perpendicular to the shore. This zigzag pattern carries water down the beach as a longshore current. Illustration by Don Vierstra.

Lesson 12/Oceans and Coastlines

Longshore currents can also be created by a process called wave setup (**Figure 12.6**). This occurs where the waves approach the shoreline at a right angle. In this case, there is no zigzag pattern to carry the water down the shoreline; rather, the water piles up along the shoreline within the surf zone. Since water is fluid, it slides off the pile in opposite directions parallel to the shoreline, creating divergent longshore currents. In general, the higher the incoming breakers are, the greater the amount of water contained within these piles of water, and the more powerful the subsequent divergent longshore currents are.

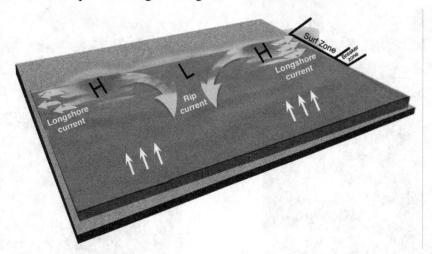

Figure 12.6 Wave Setup Effects. Longshore currents can also be created through wave setup, where the waves approach the shoreline at right angles. Water piles up in the surf zone, and then slides down the pile on either side to create a longshore current. Illustration by Marie Hulett.

Where two divergent longshore currents converge, the water flow will be diverted seaward to create a very strong rip current that transports the excess water that has been brought into the surf zone back to sea. This situation generally prompts warnings from local officials to avoid going into the water.

Waves and longshore currents are responsible for many of the erosional and depositional features seen along coastlines. If the shoreline is composed of rock, the energy from waves works to erode it. If the shoreline is composed of sediments, waves may pick them up or they may deposit more. Breaking waves can move large amounts of sediment that are then carried down the shoreline by longshore currents. Sediments near the wave base don't move as far, but form ripples that are perpendicular to the direction of wave motion.

Coastal Processes
No two coastlines are exactly alike. They exhibit differences in structure, rock types, or process. There are two basic types of coastline: (1) the emergent or high-energy coastline typified by most of North America's Pacific coast and portions of the northeast Atlantic coast; and (2) the submergent or low-energy coastline found along most of the Atlantic and Gulf states.

Emergent or High-Energy Coastlines
Point Lobos, California, is not only an exceptionally scenic portion of the California coastline but also an example of an emergent, high-energy coastline (see **Figure 12.7**).

Figure 12.7 Emergent Coastlines. An example of an emergent, high-energy coastline is Point Lobos, California, where waves pound the rocks creating erosional features. From *Planet Earth* by John J. Renton. Copyright © 2002 by John J. Renton. Reprinted by permission of Kendall Hunt Publishing Company.

Emergent coastlines are associated with active margins, that is, areas where the edge of a continent is located near an active plate boundary. The scene at Point Lobos is repeated along much of the Pacific coast, where the continental edge is rapidly rising because of the tectonic activity occurring near that coastline. Because the offshore slope along emergent coastlines is relatively steep, waves touch bottom close to the shore, the wave height builds rapidly, and little wave energy is dissipated before the waters reach the surf zone. Consequently, the waves pound against the rocks within the surf zone.

Each wave pounds the rock like a hammer, driving water under great pressure into fractures where it breaks down the rock and carries away the particles. This process undercuts the rock face. In some places, the fracture systems can form sea caves that further undermine the structure. Eventually, portions of the structure collapse into the sea, creating a wave-cut cliff (see **Figure 12.8**).

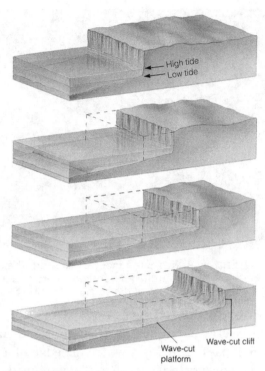

Figure 12.8 Wave-Cut Cliff and Platform. The wave-cut cliff retreats as the energy provided by the surf erodes and undermines the cliff base. Eventually, the cliff will be driven to a point above high tide, which is the furthest transgression of storm waves. The rock below the water's surface does not retreat, and it forms a wave-cut platform. Illustration by John J. Renton.

Further wave action erodes the new cliff face and the process continues, causing the wave-cut cliff to slowly retreat landward. However, only the rock above sea level is subjected to wave erosion, so the bedrock below the water's surface stays fairly intact, forming a flat surface, called a wave-cut platform, that is exposed at low tide.

Eventually, the wave-cut cliff retreats far enough inland so that it is beyond the reach of high tide and storm waves. If uplift is occurring or if sea level drops, it will expose more bedrock into which the waves will cut a second platform below the first. The sequence will repeat itself as long as uplift is occurring along the coastline or sea level is dropping. Because uplift generally occurs in fits and starts, the result is a series of steps called marine terraces. Some of the more desirable and expensive real estate on the West Coast is situated on marine terraces.

Other erosional features form where bodies of rock, called headlands, jut out into the ocean. Here waves refract to attack the headlands from all three sides. The pounding cuts sea caves into the headlands; if the sea caves meet, the remaining rock becomes a sea arch (see **Figure 12.9**). Given even more time, the arch will collapse. The forward end of the headland will then remain as a solitary sea stack near the shore. Such features can be seen all along the California, Oregon, and Washington coastlines, recording a former, more seaward location of the coastline and standing in testimony to the incredible power of wave action along high-energy coastlines.

Figure 12.9 Sea Arch and Sea Stacks. Surrounded by the surf, the sea stacks and a sea arch at the tip of the Baja peninsula in Cabo San Lucas, Mexico, are temporary reminders of the fact that the land once extended out to sea and was removed by the relentless erosion of the surf. Photo: Shutterstock 5402035, credit csp.

Submergent or Low-Energy Coastlines

In contrast to emergent, high-energy coastlines, most Atlantic and Gulf coastlines are slowly subsiding, a situation typical of passive margins where the edge of a continent is far from a plate boundary. The subsidence occurs in response to the combined effects of crustal cooling away from the oceanic ridges and the accumulation of thousands of feet of sediment within the tectonically quiet areas that border the edge of the continent. In these regions, the coastal plain of accumulated sediments extends seaward as the long, gently sloping continental shelf. Because incoming waves touch the sea bottom relatively far from the shore, water movement and bottom erosion consume much of their energy before they reach the shoreline. As a result, the amount of energy remaining in each wave as it arrives in the surf zone is significantly less than in waves approaching the typical Pacific coastline.

The low gradient of the continental shelf exaggerates slight changes in elevation, as seen in **Figure 12.10**. This creates a series of headlands and bays that form the highly irregular coastline seen along the Atlantic and Gulf coasts.

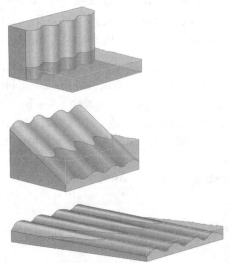

Figure 12.10 Highly Irregular Coastlines. A low gradient where the land enters the water exaggerates topographic differences, forming an irregular coastline. However, erosional and depositional processes will work to straighten it out. From *Planet Earth* by John J. Renton. Copyright © 2002 by John J. Renton. Reprinted by permission of Kendall Hunt Publishing Company.

As might be expected, the headlands are subject to erosion by wave action, causing them to slowly retreat landward. The sediment produced is carried by the longshore current to be deposited when the current's energy decreases enough that it can no longer carry the load. This sequence of events forms a number of depositional features that tend to straighten out the otherwise irregular coastline.

One area of deposition is between adjacent retreating headlands where sediments accumulate to create a bay barrier or a baymouth bar, a narrow strip of deposited sand that crosses the mouth of a bay. The water within the bay then becomes a sheltered bay or lagoon. Or the sediments may form a spit, a long, narrow deposit that is attached to the retreating headland at one end, extends parallel to the shore, and commonly terminates with a characteristic inward curl (**Figure 12.11**).

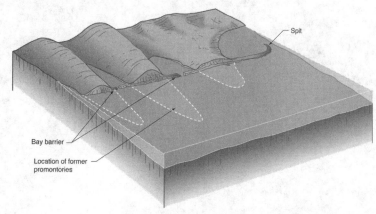

Figure 12.11 Submergent Coastline. Although submergent coastlines are initially irregular in outline, the shoreline straightens as waves erode the promontories and deposit the debris as bay barriers and spits. Illustration © Kendall Hunt Publishing Company.

In areas where there is an island close to the shoreline, the wave-deposited sediments may form a **tombolo**, a long, exposed, narrow bar of sand generally above sea level that connects the shoreline with the island (**Figure 12.12**). Tombolos also appear along the Pacific coast between the shore and sea stacks, as well as along the shoreline of some of the Great Lakes.

Figure 12.12 Tombolo. This stretch of sandbar connects St. Ninian's Isle to the coast of the Mainland, Shetland, in Scotland. Photo: Released into the public domain via Wikipedia by the photographer.

Lesson 12/Oceans and Coastlines

Perhaps the most dominant depositional feature along submergent coastlines is the many barrier islands that line the Atlantic and Gulf coasts. Barrier islands are strings of long, narrow sand islands that parallel the coastline. Between each island is a tidal inlet that allows ocean water to enter and circulate in the lagoon between the islands and the mainland. These inlets are thought to have formed when sea level rose, flooding low-lying areas behind sand ridges. Hundreds of barrier islands line the Atlantic and Gulf coast from just south of Ocean City, New Jersey, to Miami, Florida, and along the Gulf coast from the Florida Panhandle to Brownsville, Texas (**Figure 12.13**). Their presence protects the mainland from the large waves produced by storms.

Figure 12.13 Barrier Island. Photo: Courtesy of USGS.

While such depositional features are typical of submergent coastlines and erosional features are typical of emergent coastlines, both types of features are found on both types of coastline. Ultimately, local conditions determine the types of coastal features that appear in a given location.

Lab Exercises

Lab Exercise #1: *Tides*

In this laboratory exercise, you will look at the tides the way a mariner might. Below is a tide chart containing the times and heights of high and low tides over the period of one month. Tide charts are calculated and published by the National Oceanic and Atmospheric Administration (NOAA) for major commercial ports and are used by the shipping industry to ensure large commercial transport vessels can move safely in and out of a port. Boating and other ocean sport enthusiasts also rely on the charts.

Follow the instructions below and record your results in the space provided or on a separate piece of paper. Make sure to save your results. You will record your answers in the quiz at the end of this lesson.

2010 NOAA Tide Predictions: El Segundo, Santa Monica Bay

(Reference station: Los Angeles, Corrections Applied: Times: High 0 hr. 13 min., Low 0 hr. 13 min., Heights: High *0.96, Low *0.96)

January – El Segundo, Santa Monica Bay

Date	Day	Time		Height	Time		Height	Time		Height	Time		Height
01/01/2010	Fri	02:44AM	LST	1.8 L	08:59AM	LST	6.8 H	04:13PM	LST	-1.6 L	10:45PM	LST	4 H
01/02/2010	Sat	03:35AM	LST	1.7 L	09:46AM	LST	6.5 H	04:55PM	LST	-1.4 L	11:28PM	LST	4.2 H
01/03/2010	Sun	04:31AM	LST	1.6 L	10:35AM	LST	6 H	05:37PM	LST	-1.1 L			
01/04/2010	Mon	12:13AM	LST	4.4 H	05:34AM	LST	1.6 L	11:28AM	LST	5.3 H	06:19PM	LST	-0.5 L
01/05/2010	Tue	01:02AM	LST	4.6 H	06:47AM	LST	1.6 L	12:28PM	LST	4.4 H	07:03PM	LST	0.2 L
01/06/2010	Wed	01:55AM	LST	4.8 H	08:14AM	LST	1.5 L	01:44PM	LST	3.6 H	07:51PM	LST	0.9 L
01/07/2010	Thu	02:52AM	LST	4.9 H	09:51AM	LST	1.2 L	03:28PM	LST	3 H	08:46PM	LST	1.5 L
01/08/2010	Fri	03:51AM	LST	5.1 H	11:18AM	LST	0.7 L	05:27PM	LST	2.8 H	09:52PM	LST	2 L
01/09/2010	Sat	04:49AM	LST	5.3 H	12:24PM	LST	0.1 L	06:57PM	LST	3 H	11:02PM	LST	2.3 L
01/10/2010	Sun	05:42AM	LST	5.5 H	01:15PM	LST	-0.3 L	07:55PM	LST	3.3 H			
01/11/2010	Mon	12:03AM	LST	2.3 L	06:29AM	LST	5.6 H	01:56PM	LST	-0.6 L	08:36PM	LST	3.5 H
01/12/2010	Tue	12:53AM	LST	2.3 L	07:10AM	LST	5.7 H	02:32PM	LST	-0.8 L	09:08PM	LST	3.6 H
01/13/2010	Wed	01:34AM	LST	2.2 L	07:46AM	LST	5.8 H	03:03PM	LST	-0.8 L	09:35PM	LST	3.6 H
01/14/2010	Thu	02:09AM	LST	2.1 L	08:20AM	LST	5.8 H	03:32PM	LST	-0.8 L	10:01PM	LST	3.6 H
01/15/2010	Fri	02:43AM	LST	2 L	08:52AM	LST	5.7 H	03:59PM	LST	-0.7 L	10:26PM	LST	3.7 H
01/16/2010	Sat	03:16AM	LST	1.9 L	09:22AM	LST	5.5 H	04:26PM	LST	-0.5 L	10:52PM	LST	3.8 H
01/17/2010	Sun	03:51AM	LST	1.8 L	09:53AM	LST	5.2 H	04:51PM	LST	-0.3 L	11:19PM	LST	3.9 H
01/18/2010	Mon	04:28AM	LST	1.8 L	10:24AM	LST	4.8 H	05:16PM	LST	0.1 L	11:47PM	LST	4 H
01/19/2010	Tue	05:10AM	LST	1.9 L	10:57AM	LST	4.3 H	05:41PM	LST	0.5 L			
01/20/2010	Wed	12:18AM	LST	4.1 H	06:01AM	LST	1.9 L	11:34AM	LST	3.7 H	06:04PM	LST	0.9 L
01/21/2010	Thu	12:52AM	LST	4.1 H	07:07AM	LST	1.9 L	12:23PM	LST	3.2 H	06:29PM	LST	1.3 L
01/22/2010	Fri	01:34AM	LST	4.3 H	08:40AM	LST	1.7 L	01:47PM	LST	2.5 H	06:57PM	LST	1.7 L
01/23/2010	Sat	02:27AM	LST	4.4 H	10:26AM	LST	1.2 L	04:35PM	LST	2.3 H	07:43PM	LST	2.1 L
01/24/2010	Sun	03:31AM	LST	4.7 H	11:41AM	LST	0.7 L	06:39PM	LST	2.6 H	09:21PM	LST	2.4 L
01/25/2010	Mon	04:37AM	LST	5.1 H	12:33PM	LST	-0.1 L	07:25PM	LST	2.9 H	10:57PM	LST	2.4 L
01/26/2010	Tue	05:36AM	LST	5.6 H	01:16PM	LST	-0.7 L	07:58PM	LST	3.3 H			
01/27/2010	Wed	12:06AM	LST	2.2 L	06:29AM	LST	6 H	01:55PM	LST	-1.2 L	08:29PM	LST	3.6 H
01/28/2010	Thu	01:02AM	LST	1.8 L	07:19AM	LST	6.4 H	02:34PM	LST	-1.5 L	09:01PM	LST	3.9 H
01/29/2010	Fri	01:53AM	LST	1.4 L	08:06AM	LST	6.6 H	03:11PM	LST	-1.7 L	09:35PM	LST	4.2 H
01/30/2010	Sat	02:43AM	LST	1.2 L	08:53AM	LST	6.6 H	03:48PM	LST	-1.5 L	10:10PM	LST	4.6 H
01/31/2010	Sun	03:33AM	LST	0.9 L	09:40AM	LST	6.2 H	04:25PM	LST	-1.2 L	10:47PM	LST	4.8 H

All times are listed in Local Standard Time(LST) or, Local Daylight Time (LDT) (when applicable). All heights are in feet referenced to Mean Lower Low Water (MLLW).

Figure 12.14 Tide Chart. Chart by Mark Worden from data compiled by NOAA.

Instructions and Observations

The tide chart in **Figure 12.14** describes the high and low fluctuations of the tide in the Santa Monica Bay near the City of El Segundo in Southern California.

Example:

Based on the tide chart in **Figure 12.14**, January 5, 2010, had the following tide predictions:

Date: 01/05/10

Time	Height	Tide (Low or High)
1:02 a.m.	4.6 feet	High tide
6:47 a.m.	1.6 feet	Low tide
12:28 p.m.	4.4 feet	High tide
7:03 p.m.	0.2 feet	Low tide

The highest tide in the 24-hour period (on 01/05/10) = 4.6 feet

The lowest tide in the 24-hour period (on 01/05/10) = 0.2 feet

The tidal difference = 4.6 feet – 0.2 feet = 4.4 feet in a 24-hour period

Step 1: Use the tide chart (**Figure 12.14**) to record the tide predictions for January 20, 2010, in the following table. Use your recorded data to answer the questions below.

Date: 01/20/10

Time	Height	Tide (Low or High)

1. Record the highest tide in the 24-hour period.

2. Record the lowest tide in the 24-hour period.

3. Calculate the tidal difference within the 24-hour period.

4. Looking at the entire month of January, 2010, record the highest tide in the month.

5. Looking at the entire month of January, 2010, record the lowest tide in the month.

6. Calculate the tidal range for the month of January, 2010.

Step 2:   Use the information in the following scenario and the tide chart in **Figure 12.14** to answer the questions that appear below.

Elizabeth is the proud owner of a sailboat with a 42-foot mast. She wants to sail into a harbor spanned by an old bridge that has a 38-foot clearance at MHW—mean high water. Elizabeth knows two facts from her experience with this harbor:

- The harbor has approximately the same tidal highs and lows as Santa Monica Bay.
- The difference in this harbor between MHW and MLLW (mean lower low water), used in the tide chart shown in **Figure 12.14**, is about 4.7 feet.

Elizabeth knows that she can't sail under this bridge during high tide on any day and that on some days, even low tide won't provide enough clearance, especially since she likes to have at least one foot of clearance between the top of her mast and the bridge. She has to pick the right day to sail out of this harbor. But which day or days?

7. What is the bridge clearance in feet at MLLW?

8. How much clearance does Elizabeth need to sail under the bridge?

9. How low must the water height be relative to MLLW for Elizabeth to clear the bridge?

10. On which days in the first week of January, 2010, could Elizabeth plan to leave the harbor?

11. If she misses her window of opportunity, how many days will Elizabeth have to wait before she can leave?

Lab Exercise #2: *Wind-Driven Waves*

In this lab exercise, you'll use what you learned about waves to make some predictions regarding the arrival of some wind-driven waves at a coastline, something surfers, sailboarders, and scuba divers do on a regular basis. Follow the instructions below and record your results in the space provided or on a separate piece of paper. Make sure to save your results. You will record your answers in the quiz at the end of this lesson.

Instructions

Step 1:   Moderate winds about 150 nautical miles offshore have generated waves that are 3 feet high with a wavelength of about 82 feet. They are currently traveling toward land at a speed of 6 knots (nautical miles per hour). Use this information, along with what you've learned in this lesson, to answer the questions below.

12. In how many hours should you head for the beach if you want to catch a few waves from this particular storm?

13. At what water depth (mean sea level) would the wave begin to touch bottom?

14. If there were a dive boat anchored where the water was 40 feet deep at mean sea level, what, if anything, would the scuba divers exploring the sea floor be able to feel as the waves arrive?

15. If the tidal range for the location were the same as for Santa Monica Bay on 01/01/2010, would the scuba divers exploring the sea bottom feel the wave during the highest tide of the day?

16. If the divers were exploring the sea bottom during the lowest tide of the day, would they then feel the waves pass through?

17. What happens to the wave period, wavelength, and wave height as the wave approaching the shoreline reaches a depth of water equal to its wave base?

18. Waves break when the water depth is 1.3 times the wave height. If the wave breaks when it has a wave height of ten feet, how deep is the water where it breaks?

Lab Exercise #3: *Coastal Features*

In this lab, you'll use what you learned about coastal processes to identify some of the features found along coastlines. Follow the instructions below and record your results in the space provided or on a separate piece of paper. Make sure to save your results. You will record your answers in the quiz at the end of this lesson.

Instructions

Step 1: Identify the features in **Figure 12.15** by entering the name of the feature next to its corresponding letter.

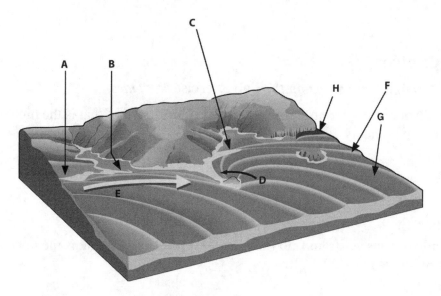

Figure 12.15 Coastal Features. Illustration by Don Vierstra.

19. Referring to the labels on **Figure 12.15** above, identify the major coastal features in the diagram from A through H.

A. _____

B. _____

C. _____

D. _____

E. _____

F. _____

G. _____

H. _____

Online Activities

Per your instructor's directions, go to the online supplement for this lab and complete the activities assigned. Viewing the online videos will help you to complete the quiz.

Quiz

Multiple Choice

Questions 1 through 9 are based on the **Lab Exercise #1:** *Tides.*

1. Record your answer from **Lab Exercise #1, Question 1**. What is the highest tide that occurred on January 20, 2010?

   a. 4.1 feet

   b. 3.7 feet

   c. 1.9 feet

   d. 0.9 feet

2. Record your answer from **Lab Exercise #1, Question 2**. What is the lowest tide that occurred on January 20, 2010?

   a. 4.1 feet

   b. 3.7 feet

   c. 1.9 feet

   d. 0.9 feet

3. Record your answer from **Lab Exercise #1, Question 3**. What is the tidal range for January 20, 2010?

   a. 5.0 feet

   b. 4.6 feet

   c. 3.2 feet

   d. 1.6 feet

4. Record your answer from **Lab Exercise #1, Question 4**. What is the highest tide for the month of January, 2010?

   a. 7.2 feet

   b. 6.8 feet

   c. 6.6 feet

   d. 5.8 feet

5. Record your answer from **Lab Exercise #1, Question 5**. What is the lowest tide for the month of January, 2010?

   a. 1.7 feet

   b. 1.9 feet

   c. −1.7 feet

   d. −0.7 feet

6. Record your answer from **Lab Exercise #1, Question 6**. What is the tidal range for the month of January, 2010?

   a. 4.5 feet

   b. 5.1 feet

   c. 6.1 feet

   d. 8.5 feet

7. Record your answer from **Lab Exercise #1, Question 9**. How low must the water height be relative to MLLW for Elizabeth to clear the bridge?

   a. –0.3 feet

   b. 1.9 feet

   c. –1.7 feet

   d. –0.7 feet

8. Record your answer from **Lab Exercise #1, Question 10**. On which days in the first week of January, 2010, could Elizabeth plan to leave the harbor?

   a. January 1, 2, 3, and 4

   b. January 2, 4, and 6

   c. January 1, 3, and 5

   d. January 2, 3, and 6

9. Record your answer from **Lab Exercise #1, Question 11**. If she misses her window of opportunity, how many days will Elizabeth have to wait before she can leave?

   a. 2 days

   b. 4 days

   c. 6 days

   d. 10 days

Questions 10 through 16 are based on the **Lab Exercise #2:** *Wind-Driven Waves.*

10. Record your answer from **Lab Exercise #2, Question 12**. In how many hours should you head for the beach if you want to catch a few waves from this particular storm?

    a. 12 hours

    b. 15 hours

    c. 21 hours

    d. 25 hours

11. Record your answer from **Lab Exercise #2, Question 13**. At what water depth (mean sea level) would the wave begin to touch bottom?

    a. about 14 feet

    b. about 20 feet

    c. about 41 feet

    d. about 82 feet

12. Record your answer from **Lab Exercise #2, Question 14**. If there were a dive boat that anchored where the water was 40 feet deep at mean sea level, what, if anything, would the scuba divers exploring the sea floor be able to feel as the waves arrive?

    a. They are above the wave base and so would feel a circular movement.

    b. They are below the wave base and so would feel a circular movement.

    c. They are above the wave base and so would not feel the waves pass through.

    d. They are below the wave base and so would not feel the waves pass through.

13. Record your answer from **Lab Exercise #2, Question 15**. If the tidal range for the location were the same as for Santa Monica Bay on 01/01/2010, would the scuba divers exploring the sea bottom feel the wave during the highest tide of the day? Why or why not?

    a. The divers are above the wave base and so would feel a circular movement.
    b. The divers are below the wave base and so would feel a circular movement.
    c. The divers are above the wave base and so would not feel the waves pass through.
    d. The divers are below the wave base and so would not feel the waves pass through.

14. Record your answer from **Lab Exercise #2, Question 16**. If the divers were exploring the sea bottom during the lowest tide of the day, would they then feel the waves pass through?

    a. The divers are above the wave base and so would feel a circular movement.
    b. The divers are below the wave base and so would feel a circular movement.
    c. The divers are above the wave base and so would not feel the waves pass through.
    d. The divers are below the wave base and so would not feel the waves pass through.

15. Record your answer from **Lab Exercise #2, Question 17**. What happens to the wave period, wavelength, and wave height as the wave approaching the shoreline reaches a depth of water equal to its wave base?

    a. The period shortens, wavelength stays the same, and wave height decreases.
    b. The period stays the same, wavelength shortens, and wave height decreases.
    c. The period stays the same, wavelength shortens, and wave height increases.
    d. The period lengthens, wavelength shortens, and wave height stays the same.

16. Record your answer from **Lab Exercise #2, Question 18**. Waves break when the water depth is 1.3 times the wave height. If the wave breaks when it has a wave height of ten feet, how deep is the water where it breaks?

    a. 1.3 feet
    b. 13 feet
    c. 26 feet
    d. 130 feet

17. Which of the following is the definition of a shoreline?

    a. the total area of interaction between land and sea
    b. the area of land between the highest and lowest possible water levels
    c. the area covered by high tide
    d. any area covered by sand

18. The height of a wind-driven wave depends upon wind velocity,

    a. duration, and fetch.
    b. duration, and distance to shore.
    c. fetch, and ocean depth.
    d. duration, and ocean depth.

19. The depth equivalent to half a wavelength is a wave's
    a. wave period.
    b. wave height.
    c. fetch.
    d. wave base.

20. Wave base determines the water depth at which a wave will
    a. break.
    b. "feel" the seafloor and begin to build.
    c. dissipate.
    d. have the most erosional power.

21. Which of the following is *most* responsible for existence of tides?
    a. the gravitational attraction of the sun
    b. the topography of the ocean bottom
    c. the gravitational attraction of the moon
    d. the rotation of Earth on its axis

22. A steady decrease in sea level following a high tide is called a(n)
    a. ebb tide.
    b. flood tide.
    c. spring tide.
    d. neap tide.

23. A steady increase in sea level following a low tide is called a(n)
    a. ebb tide.
    b. flood tide.
    c. spring tide.
    d. neap tide.

24. When the sun and moon are aligned, their combined gravitational attractions produce a(n)
    a. ebb tide.
    b. flood tide.
    c. spring tide.
    d. neap tide.

25. When the sun and moon are at right angles relative to Earth, they partially cancel out each other's gravitational attractions to produce a(n)
    a. ebb tide.
    b. flood tide.
    c. spring tide.
    d. neap tide.

26. Backwash always flows
    a. perpendicular to the slope of a shoreline.
    b. in a zigzag pattern.
    c. at the wave's angle of approach.
    d. toward a rip current.

27. Longshore currents
    a. are deep ocean currents.
    b. are caused by changes in salinity.
    c. flow in a direction parallel to the shoreline.
    d. are responsible for most erosional features.

28. Which is **NOT** true of emergent coastlines?
    a. They are often in regions being uplifted.
    b. They have very wide continental shelves.
    c. Their coastlines are dominated by erosional features.
    d. They are typical of active margins.

29. Give enough time, a sea cave is likely to erode further to become a
    a. tidal inlet.
    b. sea arch and then a sea stack.
    c. wave-cut terrace.
    d. barrier island.

30. Which is **NOT** true of submergent coastlines?
    a. Their shorelines are dominated by erosional features.
    b. They are typical of passive margins.
    c. Their continental shelves are relatively wide.
    d. Land along the coast has subsided below sea level.

31. A longshore current will deposit sediment when
    a. its energy increases.
    b. its energy decreases.
    c. its flow is disrupted by turbulence.
    d. there is a flood tide.

Short Answer

Question 1 is based on the **Lab Exercise #3:** *Coastal Features.*

1. Record your answer to **Lab Exercise #3, Question 19**. Referring to the labels on **Figure 12.15** above, identify the major coastal features in the diagram from A through H.

   A. _____

   B. _____

   C. _____

   D. _____

   E. _____

   F. _____

   G. _____

   H. _____

2. Describe the movement of a water particle as a wave passes.

3. Tides are a type of wave caused by what type of natural force between the ocean's water and the moon?

4. What are the two types of coastlines found along eastern and western coastlines of North America?

5. Name three features formed from the deposition of sediment along a coastline.

6. Name three erosional features typical of emergent coastlines.

7. What function do barrier islands serve?

GLACIERS

Lesson 13

Purpose

About 75 percent of Earth's freshwater is locked up in glacial ice, most of it in the ice sheets of Antarctica and Greenland and to a lesser extent in the "rivers of ice" known as alpine glaciers. The erosion and deposition accomplished by glaciers have sculpted much of the landscape at Earth's middle and higher latitudes. This lesson will introduce the student to glacial features and the processes that formed then.

Learning Objectives

After completing this laboratory lesson, you will be able to:
- Describe how glaciers form, and how they retreat and advance.
- Describe and identify erosional features left behind by alpine glaciers.
- Describe and identify depositional features left behind by continental and alpine glaciers.

Overview

The majority of Earth's freshwater is locked up in glaciers. The main sources of glacial ice are the ice sheets in Greenland and the Antarctic, the ice covering the Arctic Ocean, and alpine glaciers. The ice sheets, also called continental glaciers, are thick deposits of ice that form a continuous cover over a land surface with an area of more than 50,000 square kilometer (20,000 square miles). The ice sheets advance to create glacial periods, then retreat during a period known as interglacial. The last retreat occurred about 10,000 years ago, and we are currently in an interglacial period.

Alpine glaciers have been described as rivers of ice that form at the higher elevations of mountain ranges and flow down stream valleys. Alpine glaciers also increase in number and area during glacial periods. Alpine glaciers not evenly distributed throughout the world, but appear only where conditions permit. Currently most alpine glaciers are in mountains at the higher latitudes, such as the northern part of North America. However, glaciers can form at any latitude, if the temperature is cold enough. In lower latitudes, glaciers are found only above about 15,000 feet. In the mid-United States, they can form at around 9,000–10,000 feet, and near the poles, glaciers can form at sea level.

Glacier Formation

Glaciers form wherever conditions are such that snow is able to accumulate over time. When conditions remain favorable to snow accumulation, the layers can become hundreds of feet thick. Once the layers are thick and heavy enough, they respond to gravity and begin to flow downward. It is at this point that the mass becomes a glacier.

While the top layer of a glacier is snow, most of the glacier is composed of glacial ice. Glacial ice is a highly condensed version of ice that forms as the result of a long process of compression (**Figure 13.1**). Glacial ice begins as hexagonal crystals—snowflakes—accumulate as fluffy snow. The fluffiness is the result of air content, which is more than 90 percent of its volume. As layers of new snow accumulate, the ends of the crystals either break off, or they sublimate (evaporate directly into water vapor, skipping the liquid phase). At the same time, the pressure of overlying snow squeezes out the air. Through this process, the snowflakes turn into ice granules called firn, which has an air content of about 25 percent. As firn is buried even deeper under accumulating layers, increasing pressure squeezes out more air and causes the edges of the firn to begin to melt. The liquid water flows into the pore spaces, eliminating much of the remaining air. When temperatures drop, this mass of compressed firn and water recrystallizes under pressure to form glacial ice, a dense mass of interlocking crystals.

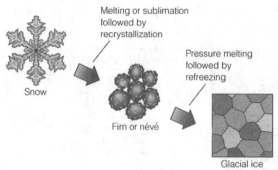

Figure 13.1 Snow to Glacial Ice. Illustration by John J. Renton.

The rate at which snow converts to glacial ice depends on the amount of accumulated snow and on climate. Warmer summers, such as those in the temperate regions, melt snow and ice more quickly; because this melting is key to the process of glacial ice formation, warmer summers facilitate the transformation from snow to glacial ice. In temperate regions, the transformation of snow to glacial ice may take just a few years. In the polar regions, however, where temperatures are constantly below the freezing point, the process may take hundreds of years.

Movement of Glacial Ice

Like other masses on Earth, glaciers are subject to the effects of gravity. The movement of glaciers is accomplished mainly through the processes of internal flow and basal slip. Internal flow is the movement of glacial ice located deeper than about 46 meters (150 feet). Under the pressure of this depth, the glacial ice behaves as a plastic material and flows under stress. In this case, the stress is gravity, which causes the ice to slowly flow downslope. The actual speed depends on both steepness of the slope and the thickness of the overlying ice. The steeper the slope and the thicker the mass of glacial ice it contains, the greater the rate of movement.

Ice less than 46 meters below the surface, however, remains brittle. Because this top layer of ice doesn't flow, the movement of the flowing ice beneath it creates tensional forces on this upper layer, which responds to the stress by fracturing (**Figure 13.2**). The fractures, called crevasses, are perpendicular to the direction of flow, and may extend all the way to the bottom of the brittle layer, called the zone of fracture.

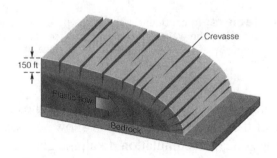 

Figure 13.2 The Zone of Fracture. Once the thickness of glacial ice exceeds 150 feet (46 meters), the lower portion of the ice begins to undergo plastic flow, applying tensional forces on the overlying brittle ice. The brittle surface ice breaks into fractures called crevasses as it is carried along by the flowing ice below. Illustration by John J. Renton; Photo: Shutterstock #59077327, credit Vladimir Melnik.

In addition to flowing downslope, alpine glaciers also slide downslope in a process called basal slip (**Figure 13.3**) in which the glacier slides down the bedrock like a child slipping down a playground slide. The steeper the slope, the faster the glacier can move. Meltwater, where it exists, helps this process along by accumulating between the bedrock and the glacier, providing lubrication that facilitates the glacier's downhill movement.

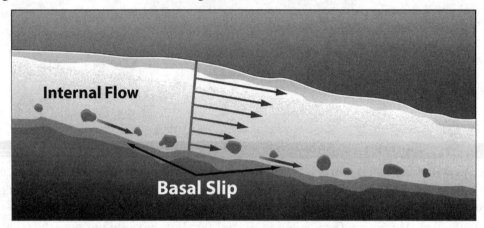

Figure 13.3 Glacial Movement. Glacial movement is a combination of basal slip and internal flow. Illustration by Don Vierstra.

Basal slip plays a major role in the movement of alpine glaciers, in which the combination of basal slip and internal flow can be as much as several meters per day. Basal slip is not typically a factor in the movement of continental glaciers, however. The temperatures in these areas are too frigid year-round for meltwater to accumulate so glacial ice remains solidly frozen to the bedrock, the force of gravity on the gentle slope insufficient to break it loose. Any movement in a continental glacier occurs instead through internal flow. Continental glaciers tend to form domes, with thicker ice in the area of formation than at the margins. The ice tends to flow from the center toward the margins similar to the way a ball of Silly Putty® spreads out when you set it out on a table (**Figure 13.4**).

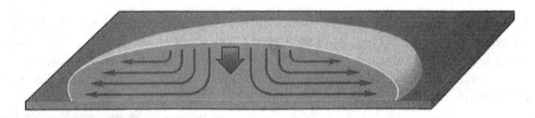

Figure 13.4 Internal Flow of Continental Glaciers. In continental glaciers, where the ice in the center of the ice mass may be thousands of feet (meters) thick, the weight of the overlying ice causes the ice mass to move downward and spread out in much the same way as pancake batter on the surface of a griddle or Silly Putty® on a table. Illustration by John J. Renton.

The Glacial Budget

For most glaciers, ice accumulates during the winter season when snowfall is the highest. Although snow may fall over the entire surface of the glacier, the buildup of snow necessary to produce glacial ice occurs in an area called the zone of accumulation. In alpine glaciers, the zone of accumulation is located at the higher elevations (**Figure 13.5**), whereas in continental glaciers it is typically located in the interior of the ice sheet.

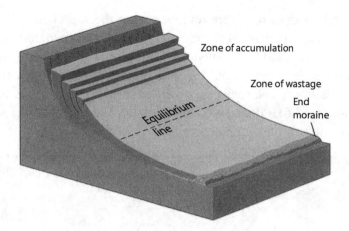

Figure 13.5 The Glacial Budget. Whether a glacier is advancing or retreating depends upon whether the mass of ice being formed in the zone of accumulation is greater or less than the amount of ice being lost in the zone of wastage. Illustration by John J. Renton.

Even as glaciers add mass in the zone of accumulation, they also lose mass, primarily in an area called the zone of wastage. In an alpine glacier, the zone of wastage is at its leading edge where the warmer temperature of the lower elevation causes the glacial ice to melt, reducing the glacier's mass. In continental glaciers, the zone of wastage is also at the edges; in this case, some of the loss is the result of sublimation but even more occurs where the ice sheet meets a body of water. Then chunks of glacial ice may break off to form icebergs in a process called calving.

In general, a glacier is continuously adding mass in the zone of accumulation and losing mass in the zone of wastage. This continuous process is called the glacial budget. The line that separates the two zones is called the equilibrium line.

Glacial ice is always moving downward, but it can appear to be stationary, to advance, or to retreat, depending on its mass. The glacier maintains a steady mass and looks stationary to the outside observer when it increases its mass in the zone of accumulation at the same rate at which it loses mass in the zone of wastage.

However, when an increase in either accumulation or wastage tips the balance, the resulting change in mass will be seen at the glacier's leading edge. When the rate of accumulation exceeds the rate of wastage, the leading edge of the glacier will advance as the glacier's mass increases. When the rate of accumulation is less than the rate of wastage, the leading edge of the glacier will retreat as the glacier's mass decreases.

It is easy then to see how a small change in the local climate can affect the mass of a glacier. If temperatures increase slightly, more melting and less accumulation will occur, causing the glacier to retreat. A decrease in temperature has the opposite effect. At the current time, most of Earth's glaciers are retreating relatively rapidly.

Glacial Erosional Processes and Features

Glaciers are some of the most powerful erosional agents on Earth. As a "river of ice," an alpine glacier has tributaries that start high in the mountains and flow downhill through stream valleys until the tributaries meet to form a larger glacier, which then continues to move downhill until the glacier either runs out of ice or reaches a body of water. During its journey, the glacier picks up soil, gravel, and rocks that it carries along as it flows.

The process by which blocks of rocks are removed from the solid bedrock is called plucking. Plucking occurs when water generated at the base of the glacier penetrates into fractures within the bedrock and freezes. The expansion of the freezing water pries the bedrock apart, loosening blocks that freeze to the glacial ice. The glacier's internal flow then pulls the blocks out of their original positions. The rocks mix with other debris within the ice to become part of the glacial mass (**Figure 13.6**). Glacial ice can pluck boulders dozens of feet in diameter from the bedrock and carry them along for many miles.

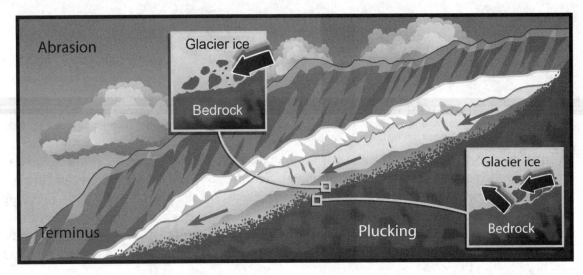

Figure 13.6 Plucking and Abrasion. Rock fragments plucked and abraded during a glacier's advance become part of the glacial mass and internal flow. The fragments that slip along the bottom often gouge striations into the bedrock. Illustration by Don Vierstra.

The combination of rock fragments and ice is extremely abrasive, not unlike a giant sheet of sandpaper. The glacier literally grinds its way through the landscape, adding to its mass the new rock particles produced by abrasion. Slowly, the glacier undercuts valley sides, only to have mass wasting complete the sculpting; this, in turn, contributes even more rock and soil to the glacial mass below the slope movement. This process repeats over many thousands of years—or even longer. When the climate changes, the glacier recedes and leaves behind dramatic changes in the contours of the bedrock. In addition, the bedrock may be polished by the abrasion, or it may be grooved with long striations that indicate the direction of glacial movement (**Figure 13.7**).

Figure 13.7 Striations. As glaciers advance, rock debris in the bottom of the ice abrades and gouges the rock surface, forming striations that allow geologists to determine the direction of glacial movement. Photo: Courtesy of P. Carrara/USGS.

Much of the spectacular mountain scenery around the world is the product of the erosional power of alpine glaciers. One common feature is the cirque, a bowl-shaped depression carved out by the ice at the former head of the glacier or glacial tributary (**Figure 13.8**). While the glacier is active, ice occupies the bottom of the cirque basin Once the ice has melted, the cirque's depression often fills with water to become a mountain lake called a tarn.

Figure 13.8 Tarn. As glacial ice melts from a cirque, the basin commonly becomes filled with water to form a scenic mountain lake called a tarn. Photo: Shutterstock 34693321, credit Herr Petroff.

Figure 13.10 Col. Erosion of a ridge where cirques meet may form a high mountain pass called a col. Photo: Courtesy of W. B. Hamilton/USGS.

Figure 13.9 Arête. Results of alpine erosion include sharp mountain ridges called arêtes. Photo: Courtesy of W. B. Hamilton/USGS.

Figure 13.11 Horn. The Matterhorn, one of the most famous horns in the world, was formed when multiple cirques carved into the bedrock around a mountain peak. Photo: Shutterstock 29014780, credit kohy.

If the glacier carves out these cirques on two sides of a ridge, the depressions begin to overlap and their upper slopes steepen. In time, the steep slopes from opposite sides of the ridge intersect along the summit of the ridge to form a knife-edged mountain ridge called an arête (**Figure 13.9**).

Some cirques will widen enough to carve through the ridge altogether. With no support from below, the overlying rock collapses to form a high mountain pass through the ridge called a col. Road builders and hikers alike commonly take advantage of cols to build roads and forge trails across glaciated mountain ridges (**Figure 13.10**).

One of the most spectacular erosional features is the horn. Horns are formed when multiple cirques are carved around a mountain peak, producing a sharp mountain spire between them. Perhaps the best-known example of a horn is the Matterhorn in the Swiss Alps (**Figure 13.11**).

From its head, the glacier flows downward. As it does, it begins to scour and abrade the floor and walls of the original V-shaped stream valley, straightening the course of the valley and widening it into a U-shape. One of the best examples of a U-shaped valley in the United States is Yosemite Valley in the California's Sierra Nevada Mountains (**Figure 13.12**). In some places, such as the Alaskan coastline, glaciers have gouged U-shaped valleys into what is now the continental shelf. When the sea level rose with the melting of the glaciers, the valley filled with water to become fjords.

Each tributary glacier also carves its own U-shaped valley, but not necessarily to the depth of the valley carved by the main glacier. After the ice retreats, the valleys carved by the tributary glaciers terminate above the main valley, seeming to hang above it. As such, they are called hanging valleys and are often sites of spectacular waterfalls, where tributary streams plunge hundreds of feet to the main valley floor.

Figure 13.12 Yosemite Valley. California's Yosemite Valley is a classic U-shaped valley; the falls to the right are tumbling from a hanging valley. Photo: Shutterstock 1666581, credit Natalia Bratslavsky.

Lesson 13/Glaciers

Within the valleys, the glaciers carve out basins during their advance and leave other depressions in the debris as they retreat. These basins and depressions often fill with water to become glacial lakes that may form strings connected by streams. Early geologists named these strings of lakes paternoster lakes, in reference to their resemblance of prayer beads.

Another erosional feature is the roche moutonnée (French for *wooly rock* or "rock sheep") (**Figure 13.13**). Roche moutonnées are elongated bedrock hills. One side is smooth and polished and slopes in the direction of the advancing glacier. The other side is blunt and has a rough surface created by plucking.

Figure 13.13 Roche Moutonnée. A roche moutonnée, like Lembert Dome in California's Yosemite National Park, is a rock mass sculpted during a glacier's advance by a combination of abrasion and plucking. Photo: Shutterstock 26241292, credit Geir Olav Lyngfjell.

Roche moutonnées are also formed by the advance of continental glaciers. In general, however, the topography left behind by continental glaciers is somewhat more subtle than that of alpine glaciers, characterized by gently rolling hills and broad flat expanses. The flatter topography left behind by a continental glacier is more the product of its retreat than of its advance. Continental glaciers do grind through hills and valleys, but erode the surfaces in equal amounts so that the relief of the region remains essentially the same. When a glacier retreats, however, it drops large amounts of debris that fill the valleys. This flattens the landscape, leaving behind only gently rolling hills such as those in Ohio and Indiana.

Glacial Depositional Processes and Features

Eventually, glaciers drop the rock material they carry. Some deposits are made when the glacial budget is balanced and the leading edge is neither advancing nor retreating. Then rock material moving along as part of the glacier's internal flow drops out of the glacier when the ice holding it melts at the leading edge.

Other glacial deposits are made as the glacier retreats, the particles dropping out as sediments as the ice melts. Unlike the deposits made by running streams, however, a glacier drops sands, silts, and boulders in no particular order; when the ice holding them melts, they fall to the ground. Thus, boulders, sand-, and clay-sized particles are often mixed in the same, poorly sorted deposit. Some of these particles may be subsequently carried away by glacial meltwater.

The sediments deposited by a glacier are collectively called drift. There are two types of drift—till and stratified drift. Till is a deposit dropped by the glacier as it retreats. It is poorly stratified, for the reasons described above (**Figure 13.14**). Over time, weathered till deposits may become overgrown with vegetation.

Figure 13.14 Glacial Till. Glacial till of the Harbor Hill terminal moraine exposed on Long Island, New York. Photo: Courtesy of USGS.

Stratified drift is sediment that was carried away from the glacier by meltwater streams and later deposited downstream with the particles settling out in the same order they would with any other stream: largest particles first, silts and clays last. In this case, the drift is stratified, or layered, with the larger particles in the lower layers and the finer particles in the top layers of each individual deposit.

A glacier in retreat produces large deposits of till and less of stratified drift. However, the retreat is rarely orderly; the leading edge advances or retreats as climate dictates. Consequently, the till deposit is not evenly spread out over the landscape but accumulates in piles and ridges that form some of the rolling landscape of a glacial terrain.

The most common glacial deposits are moraines, the general name given to any mound, ridge, or other distinct accumulation of till deposited directly from glacial ice. There are five types of moraines (**Figure 13.15**):

- Terminal or end moraine, which marks the farthest reach of the glacier before it recedes;
- Ground moraine, a blanket of till deposited as a glacier recedes;
- Recessional moraine, a deposit of till formed when the edge of a retreating glacier remains in one place for a long time before continued retreat;
- Lateral moraine, a deposit of till that builds along the outside edges of a glacier as it advances and is deposited as retreats; and

- Medial moraine, a ridge of till formed when two advancing glaciers meet to form a larger glacier.

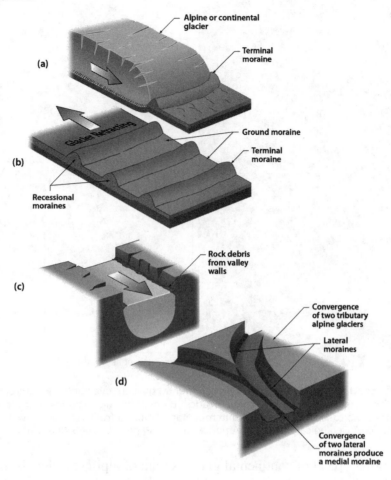

Figure 13.15 Types of Moraines. IIlustration © Kendall Hunt Publishing Company.

Other common depositional features are kettles, kettle lakes, and erratics. When blocks of ice detach from a glacier and are buried in the ground, they leave behind surface depressions, or kettles. Kettle diameters range from a few feet to several miles across. When the water table rises, some kettles become filled with water and form kettle lakes.. There are tens of thousands of kettle lakes across the parts of North America once covered by glaciers.

Much of the rock material deposited by glaciers has traveled long distances. Such rock fragments are called erratics. Erratics range in size from pebbles to boulders as large as a house, and they are often an entirely different rock type than the surrounding rocks. A distribution of many erratics in an area can often be used to determine not only the direction of ice flow, but also the location of the source rocks (**Figure 13.16**).

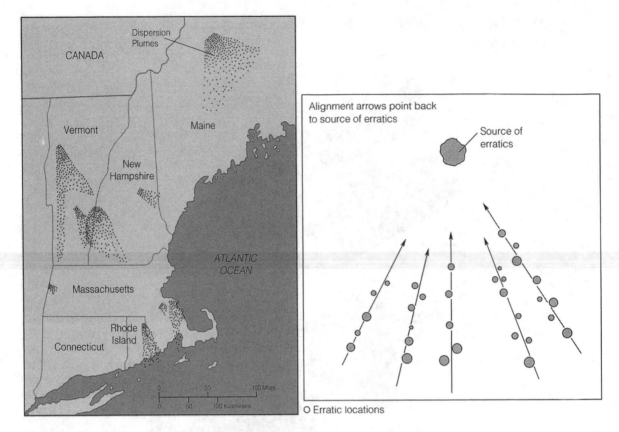

Figure 13.16 Erratic Distribution. As continental glaciers overrun rock outcrops, rock fragments can be carried off in fan-shaped patterns to become erratics. By plotting the locations of these erratics on a map, geologists are often able to locate the source outcrop. Map: Adapted from J.W. Goldthwait in R. F. Flint, "Glacial Map of North America," Geological Society of America Special Paper 60, 1945 (left); Illustration by John J. Renton (right).

A till deposit more typical of continental glaciers than of alpine glaciers is the drumlin, a low, elongate hill that looks like an inverted spoon a mile or less long and about 30 meters (100 feet) high (**Figure 13.17**). Drumlins are commonly found in groups that are oriented with the long dimension parallel with the direction of ice flow.

Figure 13.17 Drumlin. Drumlins form under advancing glaciers and align with the direction of ice movement, with the blunt end facing the upstream direction or source of the glacier. Photo: Courtesy T. Pulton/Natural Resources Canada.

Eskers are an example of a meltwater deposit. The sediment of eskers was not dropped by ice at glacier's edge but is the channel deposit of a meltwater stream that developed within a "stopped" glacier. This explains the esker's sinuous appearance (**Figure 13.18**). As running water deposits, esker deposits are well sorted in contrast to till deposits, which are poorly sorted.

Figure 13.18 Esker. Eskers are the channel deposits of streams that flowed on, through, or under the ice. Reproduced with the permission of Natural Resources Canada 2011. Photo by Jean Veillette.

Other meltwater deposits include outwash plains and valley trains. A melting glacier can produce highly turbulent streams that can pick up some of the materials in the end moraine, carrying them some distance before the stream energy decreases enough to drop them as stratified drift. In the case of a continental glacier, such a deposit is called an outwash plain; in the case of an alpine glacier, the deposit is called a valley train (**Figure 13.19**).

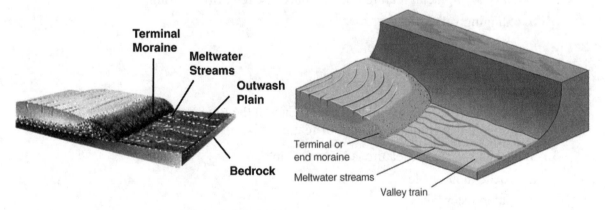

Figure 13.19 Outwash Plain and Valley Train. As the meltwater emerges from the terminus of a glacier, it carries materials that are eventually deposited as well-sorted deposits. The meltwater stream deposits in front of a continental glacier are called an outwash plain (left) and a valley train when in front of an alpine glacier. Illustrations by John J. Renton.

Lab Exercise

Lab Exercise: *Glacial Features in Yosemite Valley*

In this laboratory exercise you will apply what you have learned about glacial features and topographic maps to identify erosional and depositional features in Yosemite Valley and surrounding area. Follow the instructions below and record your results in the space provided or on a separate piece of paper. Make sure to save your results. You will record your answers to the questions in the quiz at the end of this lesson.

As you examine the contour lines on the topographic map, recall what you learned about reading topographic maps in Lesson 1, particularly the rules regarding contour lines. In particularly, keep in mind that closely spaced lines indicated a steeper slope while widely spaced lines indicate a gentler slope. This rule is essential in being able to identify the features.

Refer to the Overview for descriptions of each feature. If you have difficulty visualizing the shape of a feature based on the contour lines, drawing a topographic profile of the feature may help you to identify it. You may need to draw two topographic profiles using lines that intersect at 90° angles in order to get a complete picture of the feature's appearance.

Instructions and Observations

Step 1: Lay out the USGS topographic map of Yosemite Valley found in your lab kit.

Step 2: Find the feature called Clouds Rest on the Yosemite Valley topographic map. Examine its contours to determine what kind of feature it is.

1. What type of glacial feature is Clouds Rest?
   a. roche moutonnée
   b. hanging valley
   c. lateral moraine
   d. arête

Step 3: Find Bridalveil Creek. Examine the contours surrounding the creek to determine what type of feature it flows through.

2. What type of glacial feature does Bridalveil Creek flow through?
   a. hanging valley
   b. paternoster lakes
   c. tarn
   d. arête

Step 4: Find the feature called Liberty Cap on the topographic map and examine its contours to determine what kind of feature it is.

3. What type of glacial feature is Liberty Cap?
   a. roche moutonnée
   b. paternoster lakes
   c. tarn
   d. arête

Step 5:   On the topographic map, examine the shape of Yosemite Valley, which is itself a glacial feature.

4. What type of glacial feature is Yosemite Valley?
    a. hanging valley
    b. U-shaped valley
    c. valley train
    d. cirque

Step 6:   Two glaciers met within Yosemite Valley to form the larger glacier that carved out the valley. One glacier flowed into the valley through the Tenaya Creek stream channel and the other flowed down the Merced River stream channel. Find these two streams on your topographic map, then answer the following question.

5. What type of glacial feature would you expect to find where the two glaciers met?
    a. terminal moraine
    b. valley train
    c. medial moraine
    d. col

Examine the map to see whether there is evidence of the glacial feature that would form at the junction of the two smaller glaciers.

6. Is there evidence of this feature? If so, what landmark is this feature near?
    a. Yes, this feature is near Yosemite Village.
    b. Yes, this feature is near the Trail Center at Happy Isles.
    c. Yes, this feature is near Clark's Bridge.
    d. No, there is no evidence of this feature on the topographic map.

Step 7:   **Figure 13.20** shows what Yosemite Valley is thought to have looked like during the most extreme glaciation. **Figure 13.21** is an illustration of Yosemite Valley as it appears today. Comparing the two illustrations, it appears that the glacier filled the valley approximately to the level of the Glacier Point lookout.

Figure 13.20 Yosemite Valley During Extreme Glaciation. Illustration: Courtesy of USGS.

Lesson 13/Glaciers   **301**

RS—Rockslides	BO—Basket Dome	B — Mount Broderick
RF—Ribbon Fall	MW—Mount Watkins	SD—Sentinel Dome
EC—El Capitan	E — Echo Peaks	G — Glacier Point
TB—Three Brothers	C — Clouds Rest	SR—Sentinel Rock
EP—Eagle Peak	SM—Sunrise Mountain	FS—Fissures
YF—Yosemite Falls	Q — Quarter Domes	T — Taft Point
YV—Yosemite Village	HD—Half Dome	CS—Cathedral Spires
IC—Indian Creek	M — Mount Maclure	CR—Cathedral Rocks
R — Royal Arches	L — Mount Lyell	BV—Bridalveil Fall
W —Washington Column	F — Mount Florence	LT—Leaning Tower
ML—Mirror Lake	LY—Little Yosemite	MR—Merced River
ND—North Dome	LC—Liberty Cap	

Figure 13.21 Bird's-eye View of Yosemite Valley. This sketch of Yosemite Valley includes abbreviations that identify selected landforms, as indicated in the accompanying legend. Illustration by Natalie Weiskal/USGS.

Assuming that the elevation of the most extensive glaciation was approximately that of Glacier Point, calculate the depth of the glacier. To do this, examine the topographic map to find the elevation of the glacier's top surface at Glacier Point. Use the elevation of the benchmark (designated VABM) near the lookout.

7. What is the apparent elevation in feet above mean sea level of the glacier's surface at Glacier Point during the most extreme glaciation?

   a. 4600 feet

   b. 7214 feet

   c. 7400 feet

   d. 8122 feet

When this glacier receded, a moraine blocked the exit to Yosemite Valley, which filled with meltwater to become Yosemite Lake. Over time, the lake filled with sediment and disappeared. The depth of the sediment beneath today's valley floor is 2,000 feet.

Using your answer to Question 7 and the current elevation of the valley floor at the benchmark near Stoneman Meadow, calculate the approximate depth of the glacial mass near what is now Glacier Point during the most extreme glaciation. Recall that the glacier's base would have been at the bedrock beneath the sediments below the current valley floor.

8. How deep was the glacier near what is now Glacier Point during the most extreme glaciation?
   a. 3942 feet
   b. 4150 feet
   c. 5242 feet
   d. 6342 feet

9. How much of the depth referenced in Question 7 was glacial ice with plastic properties?
   a. 3942 feet
   b. 5092 feet
   c. 6192 feet
   d. 6296 feet

Step 8: **Figure 13.22** shows a rock that was found within Yosemite National Park. Its composition is different from the composition of the surrounding bedrock.

Figure 13.22 Erratic. Photo: Courtesy of USGS.

10. What type of glacial feature is the rock pictured in **Figure 13.22**?
    a. roche moutonnée
    b. horn
    c. drumlin
    d. erratic

Online Activities

Per your instructor's directions, go to the online supplement for this lab and complete the activities assigned. Viewing the online videos will help you to complete the quiz.

Quiz

Multiple Choice

Questions 1 through 10 are based on the **Lab Exercise:** *Glacial Features in Yosemite Valley.*

1. Record your answer to **Lab Exercise, Question 1**. What type of glacial feature is Clouds Rest?
   a. roche moutonnée
   b. hanging valley
   c. lateral moraine
   d. arête

2. Record your answer to **Lab Exercise, Question 2**. What type of glacial feature does Bridalveil Creek flow through?
   a. hanging valley
   b. paternoster lakes
   c. tarn
   d. arête

3. Record your answer to **Lab Exercise, Question 3**. What type of glacial feature is Liberty Cap?
   a. roche moutonnée
   b. paternoster lakes
   c. tarn
   d. arête

4. Record your answer to **Lab Exercise, Question 4**. What type of glacial feature is Yosemite Valley?
   a. hanging valley
   b. U-shaped valley
   c. valley train
   d. cirque

5. Record your answer to **Lab Exercise, Question 5**. What type of glacial feature would you expect to find where the two glaciers met?
   a. terminal moraine
   b. valley train
   c. medial moraine
   d. col

6. Record your answer to **Lab Exercise, Question 6**. Is there evidence of this feature? If so, what landmark is this feature near?

   a. Yes, this feature is near Yosemite Village.

   b. Yes, this feature is near the Trail Center at Happy Isles.

   c. Yes, this feature is near Clark's Bridge.

   d. No, there is no evidence of this feature on the topographic map.

7. Record your answer to **Lab Exercise, Question 7**. What is the apparent elevation in feet above mean sea level of the glacier's surface at Glacier Point during the most extreme glaciation?

   a. 4600 feet

   b. 7214 feet

   c. 7400 feet

   d. 8122 feet

8. Record your answer to **Lab Exercise, Question 8**. How deep was the glacier near what is now Glacier Point during the most extreme glaciation?

   a. 3942 feet

   b. 4150 feet

   c. 5242 feet

   d. 6342 feet

9. Record your answer to **Lab Exercise, Question 9**. How much of the depth referenced in Question 7 was glacial ice with plastic properties?

   a. 3942 feet

   b. 5092 feet

   c. 6192 feet

   d. 6296 feet

10. Record your answer to **Lab Exercise, Question 10**. What type of glacial feature is the rock pictured in **Figure 13.22**?

    a. roche moutonnée

    b. horn

    c. drumlin

    d. erratic

11. How much of Earth's freshwater is locked up in glacial ice?

    a. about 25 percent

    b. about 40 percent

    c. about 60 percent

    d. about 75 percent

12. The depth beneath the glacier's surface at which ice begins to take on plastic properties is
    a. 50 feet.
    b. 150 feet.
    c. 500 feet.
    d. 1,000 feet.

13. A shared characteristic of most glacial deposits is that they are
    a. extremely fine-grained.
    b. poorly sorted.
    c. very thick.
    d. found only in cold climates.

14. The process by which a solid may convert to a gas without going through a liquid phase is called
    a. solifluction.
    b. calving.
    c. sublimation.
    d. recrystallization.

15. Most glacial deposits are composed of
    a. till.
    b. firn.
    c. stratified drift.
    d. erratics.

16. Which of the following is **NOT** true of glaciers?
    a. They originate on land.
    b. They exist only in the Northern Hemisphere.
    c. They leave behind a changed landscape.
    d. They move large amounts of rock and sediment.

17. Which of the moraines listed below is **NOT** characteristic of continental glaciation?
    a. lateral moraine
    b. ground moraine
    c. terminal moraine
    d. recessional moraine

18. A _____ is likely to host a waterfall.
    a. horn peak
    b. paternoster
    c. calving
    d. hanging valley

19. A "bowl-shaped mountain depression" describes a(n)
    a. arête.
    b. cirque.
    c. horn.
    d. roche moutonnée.

20. Crevasses appear
    a. in glacial ice.
    b. due to compressional stress.
    c. in the zone of fracture.
    d. only as a glacier retreats.

21. A glacier appears to advance when
    a. there is more mass accumulating than is being lost.
    b. there is more mass lost than is accumulating.
    c. when the amount of mass accumulating and being lost is about the same.
    d. when there has been no accumulation or loss for several years.

22. An advancing glacier mostly creates _____, whereas a retreating glacier mostly creates _____.
    a. depositional features; erosional features
    b. erosional features; depositional features
    c. moraines; erosional features
    d. depositional features; outwash plains

23. The term *plucking* refers to
    a. glacial abrasion.
    b. a glacier depositing till.
    c. the manner in which glacial ice removes blocks of bedrock.
    d. an uneven glacial flow.

24. An esker forms when
    a. an internal stream in the glacier drops sediment.
    b. a glacier plucks blocks from one side of a rock body and polishes the other.
    c. the glacier retreats, stops, then resumes its retreat.
    d. two cirques meet.

25. An esker differs from a moraine in that
    a. it is deposited as a glacier retreats.
    b. it is sinuous and composed of stratified drift.
    c. it is a poorly sorted deposit.
    d. an esker is the result of an erosional process.

26. Continental glaciers move
   a. rarely as ice sheets are present in areas too cold to melt ice.
   b. through internal flow in all directions from a central dome.
   c. through a combination of internal flow and basal slip.
   d. primarily through basal sip.

Short Answer

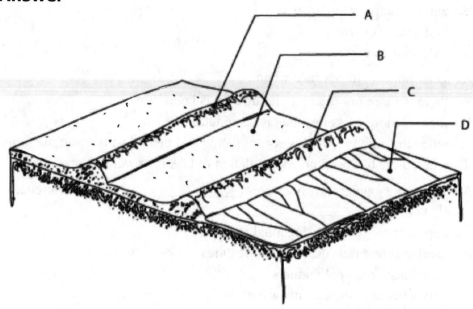

Figure 13.23 Glacial Moraine. Illustration by Mark Worden.

1. Identify the types of features shown in **Figure 13.23**, which were left behind by a continental glacier.

   a. _____

   b. _____

   c. _____

   d. _____

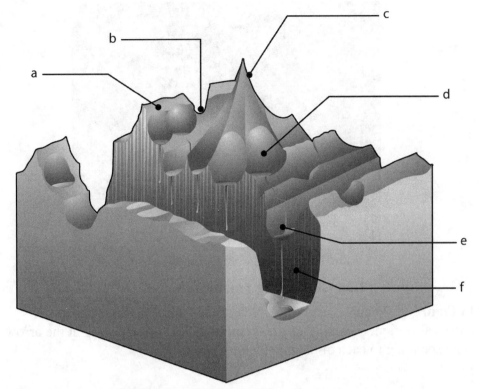

Figure 13.24 Glacial Features of a Moraine. Illustration by Don Vierstra.

2. **Figure 13.24** above depicts different types of erosional features left behind by an alpine glacier. Identify each feature in the diagram from a through f.

a. _____

b. _____

c. _____

d. _____

e. _____

f. _____

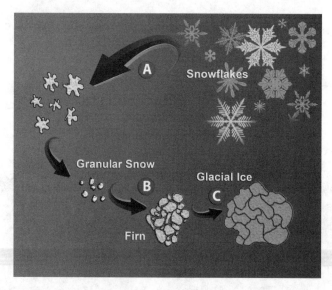

Figure 13.25 Glacial Ice Formation. Illustration by Marie Hulett.

3. In **Figure 13.25** above, identify which of the following three processes—recrystallization, accumulation, and compaction—is happening at the arrow corresponding to each of the three letters.

   a. _____

   b. _____

   c. _____

4. How do plucking and abrasion grind through a landscape?

5. What are striations, and how do they form?

6. Explain why glacial till is poorly sorted.

ECONOMIC GEOLOGY AND RESOURCES

Lesson 14

AT A GLANCE

Purpose

Learning Objectives

Materials Needed

Overview

Lab Exercises

Online Activities

Quiz

Purpose

This laboratory lesson will familiarize you with the procedures and techniques used to interpret subsurface geology as it applies to the exploration for and production of oil and natural gas.

Learning Objectives

After completing this laboratory lesson, you will be able to:

- Explain how oil and natural gas form.
- Describe structural and stratigraphic traps that trap oil and natural gas.
- Draw and interpret subsurface contour maps.

Materials Needed

- ❏ Pencil
- ❏ Eraser
- ❏ Calculator

Overview

Hydrocarbons are substances composed primarily of hydrogen and carbon, elements that readily combine with oxygen, including during combustion. Hydrocarbons burn, releasing energy as they do. The hydrogen and carbon in petroleum and coal were part of organic matter deposited millions of years ago that transformed over time to their present state. As such, oil and natural gas—collectively known as petroleum—and coal are often called fossil fuels, even though petroleum contains no fossils. Coal, however, does often contain fossils of the plant materials from which it formed. The burning of fossil fuels provides most of the world's energy for heat, electricity, and transportation.

The world's first commercial oil well was drilled in 1858 in Oil Springs, Ontario, Canada. The following year, E. L. Drake completed the first oil well drilled in the United States solely for the production of oil near Titusville, Pennsylvania. The major use of oil at the time was to provide kerosene for lamps, replacing the much more expensive oil obtained from whales. The ascendency of oil as the major energy source was the result of the invention of the mass produced, gasoline-powered Model T automobile by Henry Ford in 1908. By 1910, oil was the major source of energy in the United States. Today, oil and natural gas provide about 70 percent of the U.S. energy budget.

The rise of petroleum as a major source of fuel began a search for oil resources that continues to this day. The early wildcatter who could find oil truly found "black gold." Oil was—and still is—the ticket to great wealth and power, and so petroleum prospecting in the early twentieth century drew the same types of fortune-seekers as the earlier gold rushes. The first prospectors looked for oil seeps and drilled there. They began to have more success when geologists and others figured out how oil formed and accumulated, giving them clues as to where pools of oil might exist.

Oil and Natural Gas Formation

Oil is typically a marine product that forms after the remains of tiny aquatic and marine organisms, particularly plankton, sink to the deep seafloor where there is little oxygen (**Figure 14.1**). Some lakes have also provided suitable conditions for oil formation. Either way, the organic remains do not decompose in the absence of oxygen, and so they accumulate along with silts and clays, the clasts that form shale. Indeed, oil is frequently found in shale, and many shale formations contain large amounts of oil. Over time, the organic remains that exist between the grains of other sediments transform into a solid called kerogen. In some cases, the process stops at this point. In other cases, the layers are buried deeply enough for temperature to rise to a point where the heat begins to convert the kerogen, first into liquid oil and later, as temperatures continue to rise, into gaseous natural gas. The actual mechanisms and chemical reactions involved are not yet fully understood.

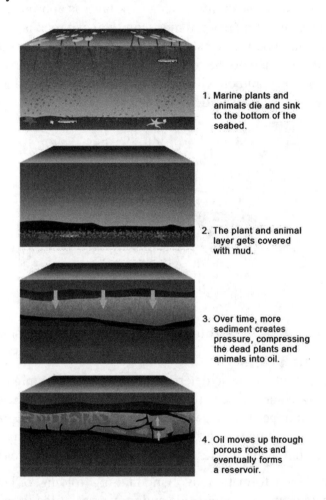

1. Marine plants and animals die and sink to the bottom of the seabed.

2. The plant and animal layer gets covered with mud.

3. Over time, more sediment creates pressure, compressing the dead plants and animals into oil.

4. Oil moves up through porous rocks and eventually forms a reservoir.

Figure 14.1 Oil and Natural Gas Formation. The formation of petroleum begins (1) when tiny organisms die and sink to the ocean or lake bottom, an oxygen-poor environment where they accumulate (2) along with other sediments. Over time, the pressure from overlying layers compresses the sediments (3), and the organic materials convert to kerogen. If buried deeply enough, the temperature may rise enough to convert the kerogen to oil and natural gas. Oil and natural gas are buoyant enough to rise through the rock (4). If a rock is porous, it becomes a reservoir rock that stores the oil and gas. In addition, overlying impermeable rock layers or other structures may trap the petroleum, storing it indefinitely. Illustration: Courtesy of Library and Archives Canada.

Over time, the weight of accumulating overlying layers of rock literally squeezes the natural gas and oil out of the sedimentary layer in which they formed. These fluid substances ooze into overlying layers until they find a reservoir rock that is both permeable and porous, the same type of rock that composes aquifers. Like aquifers, most petroleum reservoirs are sedimentary rocks, with sandstones being the best reservoir. Limestone can also be an excellent reservoir because of the porosity and permeability of the rock and the frequent presence of cavernous openings. As one might expect, rocks with low permeability are poor reservoirs, even though some, like shale, can contain large amounts of oil as a result of high porosity.

The journey of natural gas and oil molecules doesn't stop when they reach a suitable reservoir rock. Because they are more buoyant than the water that may be present in the rock's pores, the gas and oil rise toward the top of the reservoir rock. If there are no obstructions, the gas and oil will reach the surface and leak out as natural tar seeps.

In most cases, however, natural gas and oil are obstructed before they reach the surface by an overlying rock with very low permeability. Such a rock layer is appropriately called a cap rock. It should come as no surprise to find that the most common cap rock is shale. The cap rock creates a petroleum trap, an impermeable structure that prevents further upward movement of natural gas and oil, confining them to the reservoir rock (**Figure 14.2**). Natural gas and oil in petroleum traps have much in common with water trapped in a confined aquifer, including the force with which they gush when released.

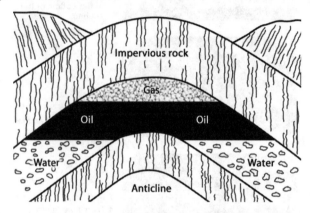

Figure 14.2 Petroleum Trap. Illustration by Bob Dixon.

Types of Petroleum Traps

The task for early oil prospectors was to determine what formations might be acting as petroleum traps. A logical structural feature is the anticline, and, indeed, much oil was found by drilling into anticlines topped with impermeable cap rock. Early geological mapping provided more clues, like certain types of faulting where similar traps could exist. Today's oil geologists use far more sophisticated tools, but the principles are the same, based on an area's geological history.

Traps are usually formed by deformation involving faulting, folding, or both. However, some traps are formed by changes in the composition and properties of the reservoir rock that affect porosity and permeability; for instance, a section of the rock might have a finer grain more closely packed together because of the manner in which it lithified. Or the reservoir rock might part of a formation of tilted strata that meets an unconformity. Some traps are the result of complex combinations of structural and stratigraphic variations that are difficult to unravel and evaluate.

There are two major types of hydrocarbon traps: structural traps, in which the deformation of the reservoir rock traps oil and gas, and stratigraphic traps, which form when permeability of the reservoir rock changes or when a tilted layer of reservoir rock is sealed off at the updip end (the high end) by another bed. **Figures 14.3**, **14.4**, **14.5**, and **14.6** provide explanations for the most common types of petroleum traps.

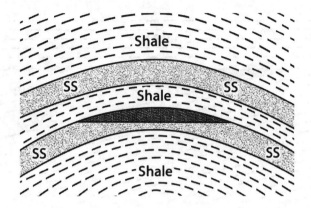

Figure 14.3 Fold-Related Structural Trap. The strata are deformed into an anticlinal fold, causing the hydrocarbons to migrate updip from both sides and accumulate in the reservoir rock at the top of the anticline. In order for petroleum to remain in the reservoir, the rock must be overlain by a low-permeability rock layer, appropriately called a cap rock. It should come as no surprise to find that the most common cap rock is shale. Illustration by Bob Dixon.

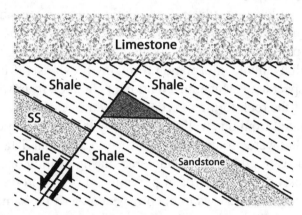

Figure 14.4 Fault-Related Structural Trap. The strata are tilted and faulted, so that hydrocarbons that migrate updip to accumulate where the reservoir rock meets a displaced impermeable rock on the other side of the fault. Structural traps are far more common than stratigraphic traps, probably because the former are easier to find than the latter. Illustration by Bob Dixon.

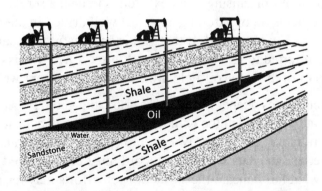

Figure 14.5 Simple Stratigraphic Trap. In this case, the hydrocarbons are trapped where two strata of impermeable rock meet to "pinch out" the reservoir rock until it tapers off completely. This is sometimes referred to as "shaling out" of the reservoir rock when the cap rock is shale. In this illustration, the pinch-out is in the updip direction, and hydrocarbons have migrated updip to the right accumulate within the pinched out portion of the reservoir rock. Illustration by Bob Dixon.

Lesson 14/Economic Geology and Resources

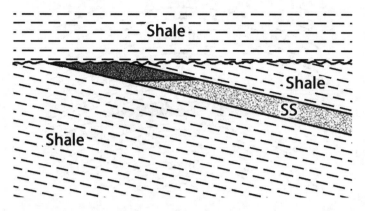

Figure 14.6 Unconformity-Related Stratigraphic Trap. In this illustration, an angular unconformity existed at the updip portion of the reservoir rocks to which the hydrocarbons migrated; the oil and gas were then trapped by the layer of shale overlying the unconformity. Illustration by Bob Dixon.

Finding Petroleum Traps

There is no direct method of locating subsurface, or below the surface, accumulations of petroleum, nor is there a physical property of underground petroleum that can be measured at the ground surface. Today's petroleum geologists located oil and gas through indirect methods. The first step is to assemble all available information from any previously drilled oil wells, gas wells, or water wells in the prospect area. Much information is obtained from well records that include descriptions and depths of the rock formations encountered as each well was drilled. The formations may be correlated from one well to another by the mineralogical content or fossil content of rock samples that were brought to the surface. Additional data might consist of information about any traces of oil or gas encountered during drilling operations, geophysical tests on any wells previously drilled, records of previous oil or gas production in the prospect area, or the results of geophysical surveys in the prospect area. If the same rock formations that occur in the subsurface are visible at the surface, the surface exposures are another source of valuable information as to what may lie underground. New visualization technologies and seismographic data, including three-dimensional mapping, can help to pinpoint areas where oil is likely to be found.

If a subsurface formation looks promising, the next step is to drill a series of test wells into the reservoir rock to see if it yields oil or natural gas. A test well drilled in the hope of discovering a new pool of oil is known as a wildcat well. If the well is successful, it is called a discovery well; if it is unsuccessful, it is referred to as a dry hole (see **Figure 14.7**). Wells drilled near the original discovery and part of the same oil field are called development wells.

As the search for petroleum goes deeper into Earth's crust, the geology becomes more complex and uncertain, and the data upon which the geologist must base a conclusion as to prospects of finding oil become less reliable. Therefore, every scrap of information must be squeezed out of the record and put to use, and the data from each record must be correlated with other data points to build up a picture of what lies beneath the surface. Such efforts may result in finding the favorable conditions that will geologically and economically justify the drilling of a wildcat well.

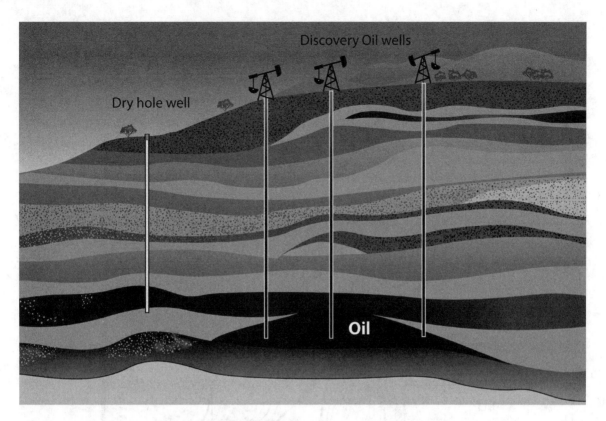

Figure 14.7 Oil Well. Illustration by Don Vierstra.

The Structure Contour Map

Part of the process of assessing a rock formation for the prospect of yielding economically viable resources, including petroleum, is to map the structure. There are two types of maps commonly used for this purpose: the contour map, which shows the depth and shape of a reservoir rock, and the isopach map, which depicts thickness. Together, these two maps can provide a three-dimensional picture of the formation in question.

A structure contour map (**Figure 14.8**) is constructed the same way as the topographic maps discussed in Lesson 1; the main difference is that the control points, which describe the top surface of the rock structure, are actually below Earth's surface. These elevations are determined through the drilling of wells and boreholes. A structure contour map, therefore, shows the structurally high parts and low parts of a prospect formation's upper surface. It describes the surface you would see looking down at a particular structure if all the overlying sediments and rock layers were removed.

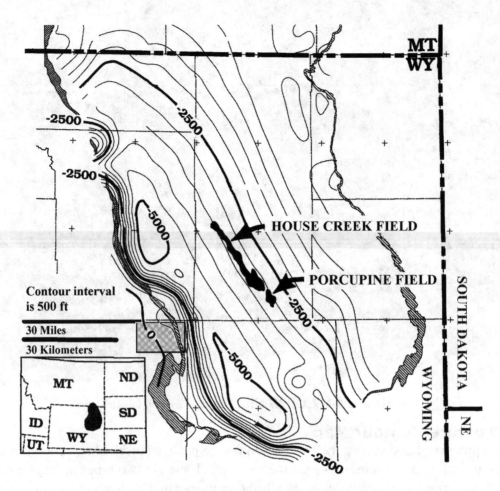

Figure 14.8 Structure Contour. Map: Courtesy of USGS.

Figure 14.8 is a structure contour for Sussex Sandstone formation in the Powder River Basin area of the area of Wyoming and Montana, an area with productive oil fields. Note that the elevations are given relative to mean sea level. Elevations below sea level are negative numbers; for instance, –2500 feet. The surface elevation of this region of the Powder River Basin is 4,554 feet.

To practice reading a structure contour map, answer the following questions about **Figure 14.8**. Answers appear in the student answer key at the end of this lesson.

1. What is the contour interval?

2. What is the index contour interval?

3. What is the elevation of the lowest index contour line?

4. What is the elevation of the lowest contour line?

5. What is the elevation of the highest index contour line?

6. What is the elevation of the highest contour line?

7. In which direction does the formation dip?

8. Within what elevation range are the oil fields?

9. Approximately how deep beneath the surface are the oil fields?

10. Why do you think the oil fields are not located at the highest elevation in the formation?

In fact, there *is* a stratigraphic trap in the Sussex Sandstone. The properties of the formation sandstone changes from a porous and permeable sandstone to a relatively impermeable mudstone that keeps the oil from migrating any further updip. In addition, a chert-pebble sandstone forms a discontinuous cap over the trap.

The Isopach Map

An isopach map depicts what a structure map cannot: the thickness of a structure or rock unit. An isopach map is also a contour map, but in this case, the contour lines connect points of equal thickness rather than points of equal elevation. **Figure 14.9** is an isopach map of the White Rim Standstone, one layer of the colorful horizontal rock strata seen in Utah's Canyonlands National Park. This particular isopach map shows the thickness of the sandstone near the junction of the Green River and Colorado River.

It's important to remember that the numbers on the contours represent thickness and not the elevation or depth of the White Rim Sandstone's surface. In addition, remember that there are other rock strata above and below the White Rock Sandstone.

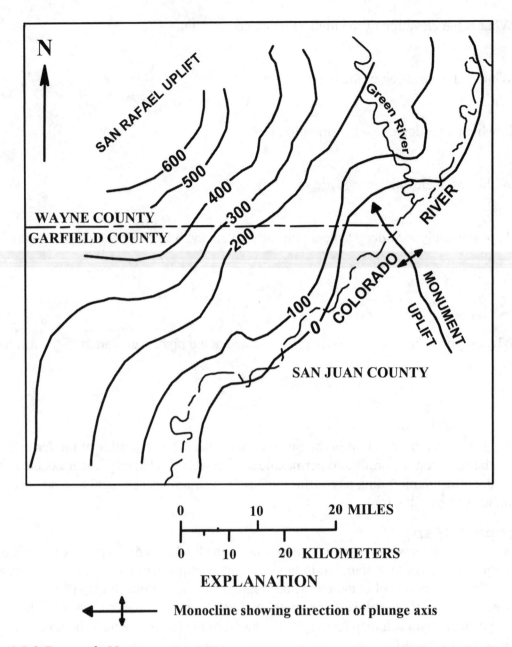

N

SAN RAFAEL UPLIFT

600
500
400
300
200

WAYNE COUNTY
GARFIELD COUNTY

Green River

RIVER

100
0 COLORADO

MONUMENT UPLIFT

SAN JUAN COUNTY

| 0 | 10 | 20 MILES |

| 0 | 10 | 20 KILOMETERS |

EXPLANATION

Monocline showing direction of plunge axis

Figure 14.9 Isopach Map. Isopach map of White Rim Sandstone. Map: Courtesy of USGS (modified from Baars and Seager, 1970).

To practice reading an isopach map, answer the following questions about **Figure 14.9**. Answers appear in the student answer key at the end of this lesson.

1. Where is the White Rim Sandstone the thickest, and what is the range of its maximum thickness?

2. Where is the White Rim Sandstone the thinnest, and how thin does it become?

3. Why do you think the isopach map showed a contour line for 0-foot thickness rather than stopping at 100 feet as the minimum thickness?

The combination of a structure contour map and an isopach map yields much about the shape of the structure. The contour map reveals whether the top of the structure is horizontal or whether it dips. The isopach reveals not only the thickness, but the contours of the bottom surface of the structure, as the elevation of the bottom surface must be the elevation of the top surface minus the thickness. For instance, if the top surface is at 2400 feet in elevation and the thickness is 200 feet, the elevation of the bottom surface must be 2200 feet. Thus, the shape of the top surface combined with thickness data reveals the entire shape of the reservoir rock.

Even so, since petroleum collects at the highest possible point in a structure, geologists are much more interested in the shape of the top surface than the bottom surface.

To grasp how the two maps work together to provide a complete picture of the shape of a reservoir rock, use the isopach map in **Figure 14.9** to answer the following questions. The answers appear in the student answer key at the back of this lesson.

4. Draw a simple side view perspective of the thickness of the portion of White Rim Sandstone seen in **Figure 14.9** with west being to the left and east being to the right. Assume that the top surface of the White Rim Sandstone is nearly horizontal.

5. What would a side view of the White Rim Sandstone look like if the top surface were steeply dipped and the bottom surface were horizontal?

6. What would a side view of the White Rim Sandstone look like if the top surface were only slightly dipped?

7. Looking at the drawings, how would a geologist describe the appearance of this formation where its thickness diminishes to 0 feet?

Interpreting Maps to Find Petroleum

Structure contour maps and isopach maps work together to help petroleum geologists identify possible locations where quantities of oil and natural gas may have accumulated. Identifying such locations depends upon several rules based both on the behavior of the migrating petroleum and on economic feasibility.

(1) Structurally high regions, such as the tops of folds (anticlines), are more likely to contain oil or gas than structurally low regions. This is because oil and gas migrate upward until trapped.

(2) The flanks of the structural highs might be considered because oil or gas could be trapped against faults, changes in lithology, or updip pinchouts at those sites.

(3) The reservoir rock must be sufficiently thick to hold enough petroleum for drilling and extraction to be economically feasible. Otherwise promising structures are often eliminated because they are too thin and can't trap enough petroleum to pay for the cost of extracting it.

Lab Exercises

Lab Exercise #1: *Structure Contour Map*

In these exercises, you will play the role of petroleum geologist to work with structure and isopach maps in order to identify where oil may be found in subsurface strata. Follow the steps below and record your observations in the space provided or on a separate piece of paper. Make sure to save the results. You will record your answers in the quiz at the end of this lesson.

Instructions and Observations

Step 1: Locate **Figure 14.10**, a structure contour map of a reservoir rock containing an oil field based on the depth of wells that have been drilled into it. The numbers represent the subsurface elevations of the wells relative to mean sea level. Assume the elevation of Earth's surface at this location is 7500 feet.

Step 2: Using a 50-foot contour interval, draw the contours of the map on **Figure 14.10**. Adjustments to your drawing are almost inevitable, so use a pencil. You might also want to photocopy the exercise or download a copy of this exercise from the online supplement for this lab to you can have a fresh copy for each attempt.

Step 3: As indicated on the map legend, open circles represent dry holes and the black dots represent oil-producing wells. Use the structure contours on the map to assist you in identifying the approximate location of the oil boundary. Use a dashed line to mark this area.

Step 4: Shade the area that contains the oil-producing wells on the structure contour map. Submit your work as directed by your instructor.

Step 5: Answer the following questions regarding the contour map you just drew.

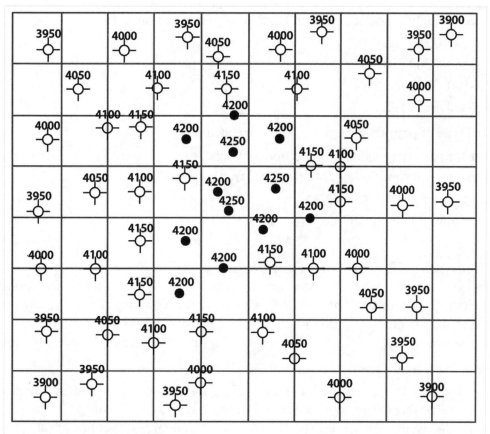

Contour Interval = 50 ft.

⊖ = Dry holes - not oil producing wells, may contain water within the well.

● = Oil well - oil producing well.

Figure 14.10 Structure Contour Map. Illustration by Bob Dixon.

1. What is the elevation of the highest contour line?

2. What is the elevation of the lowest contour line?

3. At what elevations are the productive oil wells?
   a. 4000 and 4150 feet
   b. 4100 and 4150 feet
   c. 4150 and 4200 feet
   d. 4200 and 4250 feet

4. Within what elevation range is the oil field?
   a. 4000–4300 feet
   b. 4200–4250 feet
   c. 4150–4300 feet
   d. 4000–4250 feet

5. Approximately how deep beneath Earth's surface is the oil field?

 a. 3200–3350 feet

 b. 3250–3300 feet

 c. 3100–3500 feet

 d. 4000–4250 feet

6. What type of petroleum trap does this appear to be?

 a. a simple stratigraphic trap (updip pinchout)

 b. an unconformity-related stratigraphic trap

 c. a fold-related structural trap

 d. a fault-related structural trap

Lab Exercise #2: *Finding Petroleum*

In this exercise, you will use existing structure contour maps overlaid with isopach contour lines to identify the best locations in which to drill test wells for oil or natural gas. Feel free to refer to the rules in the section on "Interpreting Maps to Find Petroleum" on page 322 and to the drawings of the various types of petroleum traps in **Figures 14.3 through 14.6**.

Instructions and Observations

Step 1: Locate and examine **Figure 14.11**. The figure is composed of two maps on the same sheet. The solid lines are a structural contour map of the Bonanza Sandstone, a reservoir rock that has produced large quantities of oil. The numbers on the solid contour lines represent the elevation of the top of the Bonanza Sandstone. This sandstone is buried thousands of feet below the surface, but the numbers are positive because the surface of this structure is at least 1300 feet above sea level. The structural contour map shows the shape of the top surface of the Bonanza Sandstone and tells the geologist which type of structure is located at depth. Since oil tends to migrate upward in the reservoir rock, the geologist tries to find the area where the Bonanza Sandstone is at the highest elevation.

The second map in **Figure 14.11**, shown by the dashed line, is an isopach map that indicates the thickness of the Bonanza Sandstone. In general, the thicker the reservoir rock, the more oil it can hold and yield when drilled. The dashed lines show that in the area covered by the map the Bonanza Sandstone is not present in some areas and is thick in other areas.

Step 2: Take a few moments to examine **Figure 14.11**. Remember that the solid lines represent the elevations of contour lines and that the dashed lines represent the thickness of the Bonanza Sandstone.

Reading an overlay map can be confusing. A helpful hint is to focus first on one map and then the other, being sure not to confuse the values for the dashed lines with the values for the contour lines.

- Focus on the solid lines (while ignore the dashed lines and their values) to get a picture of the contours of the top surface of the reservoir rock.

- Focus on the dashed lines (while ignoring the solid lines and their values) to get a picture of how the thickness of the reservoir rock changes over distance.

Once you understand what each map illustrates, it's easier to put the two together to form a more complete picture of the shape of the reservoir rock.

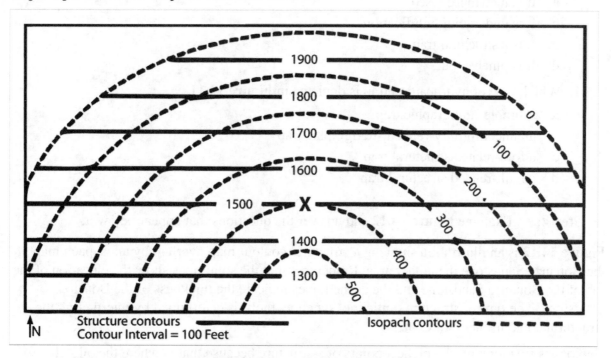

Figure 14.11. Illustration by Mark Worden.

Step 3: Answer the following questions about **Figure 14.11**.

7. What is the lowest elevation of the top surface of the reservoir rock?

8. What is the highest elevation of the top surface of the reservoir rock?

9. Does this structure dip?

10. Recalling that the contour lines represent the elevation of the *top* surface of the reservoir rock and that the isopach lines represent the thickness of the reservoir rock, what would be the elevation of the *bottom* of the Bonanza Sandstone at point X in **Figure 14.11**?
    a. 400 feet above sea level
    b. 1,500 feet above sea level
    c. 1,100 feet above sea level
    d. 1,900 feet above sea level

11. What is the thickness of the reservoir rock where the top surface is lowest?

12. What is the thickness of the reservoir rock where the top surface is highest?

13. What is the geological term for what is happening at the high end of this structure?

   a. It is metamorphosed.

   b. It is undergoing lithification.

   c. It is a structural trap.

   d. It is pinched out.

14. Which type of hydrocarbon trap is depicted in **Figure 14.11**?

   a. a simple stratigraphic trap

   b. an unconformity-related stratigraphic trap

   c. a fold-related structural trap

   d. a fault-related structural trap

Step 4:   Examine **Figure 14.12** and answer the questions that appear below it.

Figure 14.12 is an illustration showing a structural contour map overlain by an isopach map of the Bonanza Sandstone in another area. Remember that the solid lines show the elevation of the top of the Bonanza Sandstone and the dashed lines indicate the thickness of the Bonanza Sandstone. The type of structure (anticline, syncline, fault) is determined by interpreting the structural contour lines (solid lines).

Geologists first look for the highest points of a structure because that is where the oil will accumulate. They will also look for the thicker parts of the structure as they would hold more oil than the thinner parts of the structure. The greatest accumulation of oil will be in the higher points of the structure where the sandstone is thickest (as shown by the dashed lines of the isopach map).

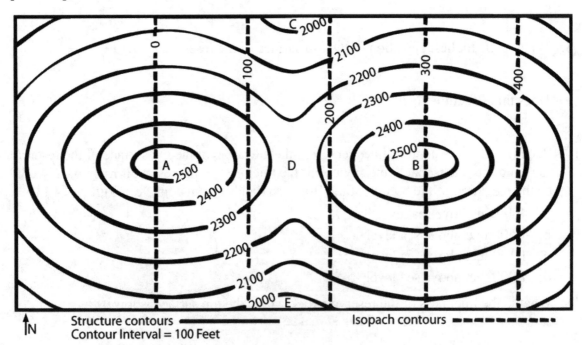

Figure 14.12. Illustration by Mark Worden.

15. What is the elevation of the highest structure contour lines?

16. What two letters are at the highest points in the structure?

17. What is the elevation range of the area around point D?

18. What kind of geologic structure has the shape depicted by this structure contour map?
    a. a joint
    b. a fold
    c. a normal fault
    d. a reverse fault

19. Describe the thickness of this formation based on the isopach contours.
    a. It is more than 400-feet thick at the lowest point of the syncline.
    b. It is more than 400-feet thick at the top of the anticline and 0 feet at the syncline.
    c. It is more than 400-feet thick to the west and pinching out to the east near the top of a fold.
    d. It is more than 400-feet thick to the east and pinching out to the west near the top of a fold.

20. The type of trap depicted in **Figure 14.12** is
    a. a simple stratigraphic trap.
    b. an unconformity-related stratigraphic trap.
    c. a fold-related structural trap.
    d. a fault-related structural trap.

Step 5: Using **Figure 14.12**, evaluate prospective locations for a test well. The answers to questions 21 through 29 appear in the student answer key at the end of this lesson to guide you in learning how geologists assess a prospective location. Questions 30, 31, and 32 will require you to apply what you've learned to make your own assessment.

21. At Point C, what is the elevation of the top of the Bonanza Sandstone?

22. What is the thickness of the sandstone?

23. Is Point C a good place to drill a test well?

24. At Point A, what is the elevation of the top of the Bonanza Sandstone?

25. What is the thickness of the sandstone at Point A?

26. Is Point A a good place to drill a test well?

27. At Point B, what is the elevation of the top of the Bonanza Sandstone?

28. What is the thickness of the Bonanza Sandstone, at Point B?

29. Is Point B a good place to drill a test well?

30. Is Point D a good place to drill? Why or why not?

31. Would you invest in drilling a well at Point E? Why or why not?

32. At which point in **Figure 14.12** would you, as a petroleum geologist, recommend for the test well to be drilled?
    a. A
    b. B
    c. C
    d. D
    e. E

Online Activities

Per your instructor's directions, go to the online supplement for this lab and complete the activities assigned. Viewing the online videos will help you to complete the quiz.

Quiz

Multiple Choice

Questions 1 through 4 are based on **Lab Exercise #1:** *Structure Contour Map.*

1. Record your answer to **Lab Exercise #1: Question 3**. At what elevations are the productive oil wells?
   a. 4000 and 4150 feet
   b. 4100 and 4150 feet
   c. 4150 and 4200 feet
   d. 4200 and 4250 feet

2. Record your answer to **Lab Exercise #1: Question 4**. Within what elevation range is the oil field?
   a. 4000–4300 feet
   b. 4200–4250 feet
   c. 4150–4300 feet
   d. 4000–4250 feet

3. Record your answer to **Lab Exercise #1: Question 5**. Approximately how deep beneath Earth's surface is the oil field?
   a. 3200–3350 feet
   b. 3250–3300 feet
   c. 3100–3500 feet
   d. 4000–4250 feet

4. Record your answer to **Lab Exercise #1: Question 6**. What type of petroleum trap does this appear to be?
   a. a simple stratigraphic trap (updip pinchout)
   b. an unconformity-related stratigraphic trap
   c. a fold-related structural trap
   d. a fault-related structural trap

Questions 5 through 11 are based on **Lab Exercise #2:** *Finding Petroleum.*

5. Record your answer to **Lab Exercise #2, Question 10**. The elevation of the *bottom* of the Bonanza Sandstone at point X in **Figure 14.11** is
   a. 400 feet above sea level.
   b. 1,500 feet above sea level.
   c. 1,100 feet above sea level.
   d. 1,900 feet above sea level.

6. Record your answer to **Lab Exercise #2, Question 13**. What is the geological term for what is happening at the high end of this structure?

    a. It is metamorphosed.

    b. It is lithifying.

    c. It is a structural trap.

    d. It is pinched out.

7. Record your answer to **Lab Exercise #2, Question 14**. Which type of hydrocarbon trap is depicted in **Figure 14.11**?

    a. a simple stratigraphic trap

    b. an unconformity-related stratigraphic trap

    c. a fold-related structural trap

    d. a fault-related structural trap

8. Record your answer to **Lab Exercise #2, Question 18**. What kind of geologic structure has the shape depicted by the structure contour map in **Figure 14.12**?

    a. a joint

    b. a fold

    c. a normal fault

    d. a reverse fault

9. Record your answer to **Lab Exercise #2, Question 19**. Describe the thickness of the formation in **Figure 14.12** based on the isopach contours.

    a. It is more than 400-feet thick at the lowest point of the syncline.

    b. It is more than 400-feet thick at the top of the anticline and 0 feet at the syncline.

    c. It is more than 400-feet thick to the west and pinching out to the east near the top of a fold.

    d. It is more than 400-feet thick to the east and pinching out to the west near the top of a fold.

10. Record your answer to **Lab Exercise #2, Question 20**. The type of trap depicted in **Figure 14.12** is

    a. a simple stratigraphic trap.

    b. an unconformity-related stratigraphic trap.

    c. a fold-related structural trap.

    d. a fault-related structural trap.

11. Record your answer to **Lab Exercise #2, Question 32**. At which point in **Figure 14.12** would you, as a petroleum geologist, recommend for the test well to be drilled?

    a. A

    b. B

    c. C

    d. D

    e. E

12. The organic matter, or hydrocarbon raw material, is derived primarily from
    a. dinosaurs.
    b. ancient swamp vegetation.
    c. mammoths and mastodons.
    d. microscopic marine and aquatic organisms.

13. During the petroleum formation process, the organic matter is preserved in an
    a. oxygenated environment with an influx of fine sediment.
    b. oxygenated environment with no influx of fine sediment.
    c. oxygen-deficient environment with an influx of fine sediment.
    d. oxygen-deficient environment with no influx of fine sediment.

14. Of the following, the best petroleum reservoir rock would be
    a. a massive granite.
    b. a banded gneiss.
    c. a massive siltstone.
    d. a poorly cemented sandstone.

15. A sedimentary rock that frequently acts as a cap rock is
    a. limestone.
    b. gypsum.
    c. sandstone.
    d. shale.

16. A sedimentary rock that often contains large amounts of oil but is not a reservoir rock is
    a. granite.
    b. gypsum.
    c. sandstone.
    d. shale.

17. A solid organic substance found in sedimentary rocks and thought to be a precursor to oil is
    a. coal.
    b. kerogen.
    c. natural gas.
    d. shale.

18. If a reservoir rock contains water, the oil and natural gas will
    a. react with the water to form a carbonate.
    b. disperse.
    c. be trapped beneath the water in the reservoir rock.
    d. rise above the water to the top of the reservoir rock.

19. _____ is when a reservoir rock terminates by tapering off between two layers of impermeable rock.
    a. Isopach
    b. Kerogen
    c. A pinch out
    d. A fold-related trap

20. The order in which fluids settle within a reservoir rock, from top to bottom, is
    a. natural gas, oil, and water.
    b. oil, natural gas, and water.
    c. natural gas, water, and oil.
    d. water, natural gas, and oil.

21. Structural traps are caused by
    a. a change in the permeability of the reservoir rock.
    b. deformation of the rock strata.
    c. pinching out of the reservoir rock.
    d. a change in the density of the reservoir rock.

22. Examples of possible structural traps are
    a. horizontal rock strata.
    b. pinch outs.
    c. angular unconformities and disconformities.
    d. folds and faults.

23. Which of the following is an example of a stratigraphic trap?
    a. an anticline
    b. an angular unconformity overlaid by shale
    c. a fault
    d. horizontal rock strata

24. The contours on an isopach map represent
    a. thickness.
    b. elevation.
    c. depth.
    d. the direction of dip.

25. Elevations on a structure contour map are generally relative to
    a. mean sea level.
    b. Earth's surface at that location.
    c. the average continental elevation.
    d. the nearest disconformity.

26. Which of the following does **NOT** give geologists data about subsurface structures?
    a. surface exposures of promising rock structures
    b. data from previously drilled wells
    c. x-ray technology
    d. seismographic data

27. If the data about a subsurface structure looks promising, the next step is to drill a
    a. discovery well.
    b. dry hole.
    c. test well.
    d. producing well.

28. A structure contour map describes
    a. Earth's surface above a structure.
    b. the top surface of a structure.
    c. the thickness of a structure.
    d. the bottom surface of a structure.

29. Oil and natural gas formation require that organic material be exposed to which of the following?
    a. lack of oxygen and low temperature
    b. lack of oxygen, pressure, and heat
    c. carbon dioxide, pressure, and heat
    d. high pressure and low temperature

Short Answer

Questions 1 and 2 are based on **Lab Exercise #2:** *Finding Petroleum.*

1. Record your answer to **Lab Exercise #2, Question 30**. Is Point D a good place to drill? Why or why not?

2. Record your answer to **Lab Exercise #2, Question 31**. Would you invest in drilling a well at Point E? Why or why not?

3. Describe the environment necessary for petroleum to form.

4. How do geologists locate oil and gas traps deep within the earth?

5. Why wouldn't an oil company be interested in oil collected within a thin layer of reservoir rock?

6. Why are there often negative numbers as values for contour lines on a structure contour map?

7. Explain the difference between a wildcat well, a discovery well, and a development well.

Student Answer Key

Pages 318–319:

The Structure Contour Map

1. What is the contour interval?

    500 feet

2. What is the index contour interval?

    2500 feet

3. What is the elevation of the lowest index contour line?

    −5000 feet

4. What is the elevation of the lowest contour line?

    −5500 feet

5. What is the elevation of the highest index contour line?

    −2500 feet

6. What is the elevation of the highest contour line?

    0 feet (mean sea level)

7. In which direction does the formation dip?

    It dips to the west.

8. Within what elevation range are the oil fields?

    −3000 to −3500 feet

9. Approximately how deep beneath the surface are the oil fields?

    7554- to 8054-feet deep. Since the oil fields are below sea level, you have to add the negative values to the 4554-foot topographic elevation of the Powder River Basin to determine the depth of the oil fields below the basin's surface.

10. Why do you think the oil fields are not located at the highest elevation in the formation?

    Ordinarily petroleum would migrate upward to the highest points in the formation. Since productive wells are located somewhat lower than that, there must be a stratigraphic trap at the elevation of the oil fields.

Pages 320–321:

The Isopach Map

1. Where is the White Rim Sandstone the thickest, and what is the range of its maximum thickness?

    The White Rim Sandstone is thickest along the contour lines in the northwest corner of the map. The maximum thickness is greater than 600 feet but less than 700 feet.

2. Where is the White Rim Sandstone the thinnest, and how thin does it become?

    The White Rim Sandstone is thinnest to the southeast near the Colorado River. There the sandstone thins to 0 feet.

3. Why do you think the isopach map showed a contour line for 0-foot thickness rather than stopping at 100 feet as the minimum thickness?

    If the contour lines stopped at 100 feet, it could indicate that the White Rim Sandstone is 100-feet thick at its edge. The fact that a 0-foot contour is shown means that the thickness gradually decreased until it disappeared.

4. Draw a simple side view perspective of the thickness of the portion of White Rim Sandstone seen in **Figure 14.8** with west being to the left and east being to the right. Assume that the top surface of the White Rim Sandstone is nearly horizontal.

5. What would a side view of the White Rim Sandstone look like if the top surface were steeply dipped and the bottom surface were horizontal?

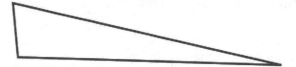

6. What would a side view of the White Rim Sandstone look like if the top surface were only slightly dipped?

7. Looking at the drawings, how would a geologist describe the appearance of this formation where its thickness diminishes to 0 feet?

    A geologist describes this as being "pinched out."

Pages 327–328:

Lab Exercise #2: *Finding Petroleum*, Step 5

21. At Point C, what is the elevation of the top of the Bonanza Sandstone?

    1900–2000 feet

22. What is the thickness of the sandstone at Point C?

    About 150 feet

23. Is Point C a good place to drill a test well?

    Point C has a good thickness of sandstone, but it is too low on the structure (anticline) and will probably produce only water.

24. At Point A, what is the elevation of the top of the Bonanza Sandstone?

    2500–2599 feet

25. What is the thickness of the sandstone at Point A?

    Just over 0 feet.

26. Is Point A a good place to drill a test well?

    Point A is poor choice. The location is over the top of the structure, which is high enough for oil to collect, but the sandstone is very thin there so it would not contain enough oil to make it economically desirable.

27. At Point B, what is the elevation of the top of the Bonanza Sandstone?

    250–299 feet

28. What is the thickness of the Bonanza Sandstone at Point B?

    About 290 feet

29. Is Point B a good place to drill a test well?

    Point B is a good place to drill because it is over the top of the structure where oil could accumulate and the sandstone is thick enough at that point for a substantial amount to be available there.

Please note: Since Questions 30, 31, and 32 are included as part of the Quiz, the answers for these questions do not appear as part of this Student Answer Key.

Lesson 14/Economic Geology and Resources

Appendix

Tables for Reference

I. Conversion Factors

Convert From	To	Multiply by
Centimeters	Feet	0.0328
Centimeters	Inches	0.394
Meters	Inches	39.37
Meters	Feet	3.2808
Meters	Yards	1.0936
Meters	Miles	0.0006214
Kilometers	Miles	0.621
Inches	Centimeters	2.54
Inches	Meters	0.0254
Feet	Centimeters	30.48
Feet	Meters	0.3048
Yards	Meters	0.9144
Miles	Kilometers	1.609
Miles	Feet	5280
Miles	Yards	1760
Square miles	Square kilometers	2.59
Square miles	Square acres	640
Grams	Ounces	0.03527
Grams	Pounds	0.002205
Kilograms	Ounces	35.27
Kilograms	Pounds	2.2046
Ounces	Grams	28.35
Pounds	Grams	453.6
Pounds	Kilograms	0.4536

II. Prefixes for System Units

Prefix	Power		Equivalent
Tera	10^{12} =	1,000,000,000,000	Trillion
Giga	10^9 =	1,000,000,000	Billion
Mega	10^6 =	1,000,000	Million
Kilo	10^3 =	1,000	Thousand
Hecto	10^2 =	100	Hundred
Deca	10^1 =	10	Ten
	10^0 =	1	One
Deci	10^{-1} =	.1	Tenth
Centi	10^{-2} =	.01	Hundredth
Milli	10^{-3} =	.001	Thousandth
Micro	10^{-6} =	.000001	Millionth
Nano	10^{-9} =	.000000001	Billionth
Pico	10^{-12} =	.000000000001	Trillionth

III. Scientific Notation

Scientific notation is a shorthand way of designating numbers and is especially useful when dealing with very large or very small numbers, both of which are difficult to read and equally difficult to notate. For example, the distance from the sun to Pluto is three billion, six hundred and sixty-seven million miles (five billion, nine hundred million kilometers). Even if we reduce the distances to numerical notation, one still has to write 3,667,000,000 miles (5,900,000,000 km). Scientific notation reduces such numbers to a number greater than 1 and less than 10, called the real constant, times 10 to an exponent. When the original number is larger than 1, the exponent of 10 is positive; when it is smaller than 1, the exponent is negative. The exponent tells how many powers of ten are needed to convert the real constant to the original number.

Let us first consider a simple case. The real constant for the number 100 would be 1. (Remember, the real constant is a number between 1 and 10.) The real constant, 1, must be multiplied by two powers of ten (10×10) to equal 100. The scientific notation for 100 would therefore be 1 (the real constant) $\times$ 10 raised to a power of two (the exponent). The scientific notation for 100 would therefore be 1×10^2. Similarly, the real constant for the number 1,000 is also 1, but it would have to be multiplied by three powers of ten ($10 \times 10 \times 10$) to equal 1,000. The scientific notation for 1,000 would therefore be 1×10^3.

Now let's convert the distance to Pluto in miles, 3,667,000,000, to scientific notation. The real constant, which would be 3.667, needs to be multiplied by nine powers of ten or 1 billion to equal the original number. The scientific notation would therefore be 3.667×10^9) miles. The scientific notation is a much easier way to both record and convey the distance from the sun to Pluto.

The same procedure is applied to determine the scientific notation of very small numbers. For example, the diameter of the carbon atom is 0.000,000,000,154 meters. The real constant would be 1.54. Because the number is less than 1, the exponent of 10 will be negative and will be equal to the number of digits the decimal point was shifted to the right to arrive at the real constant. Note that to arrive at 1.54, the decimal point is shifted 10 places. The scientific notation for the diameter of a carbon atom in meters would be 1.54×10^{-10}.

IV. Summary of the Scientific Notation Method

10^6	=	1,000,000
10^5	=	100,000
10^4	=	10,000
10^3	=	1,000
10^2	=	100
10^1	=	10
10^0	=	1
10^{-1}	=	.1
10^{-2}	=	.01
10^{-3}	=	.001
10^{-4}	=	.0001
10^{-5}	=	.00001
10^{-6}	=	.000001

V. Mineral Identification Key

Metallic Luster			
Hardness	**Streak**	**Color, etc.**	**Mineral**
6–6.5	greenish to black	brassy yellow, Specific Gravity (Sp. Gr.) = 4.2	Pyrite (iron sulfide, FeS_2)
6	dark gray	dark gray to black, submetallic luster, attracts magnet, Sp. Gr. = 5.18	Magnetite (iron oxide)
5–6.5	red to red-brown	silver to gray to brown, Sp. Gr. = 5.26	Hematite (iron oxide)
3.5–4	white to yellow-brown	brown to yellow, submetallic luster, 6 cleavage dir., Sp. Gr. = 4	Sphalerite (zinc sulfide)
2.5	gray to dark gray	steel gray, cubic cleavage, Sp. Gr. = 7.4	Galena (lead sulfide)
1	dark gray	gray to black, slippery feel, Sp. Gr. = 2.3	Graphite (carbon)

Nonmetallic Luster, Dark Color (with cleavage)			
Hardness	**Cleavage directions**	**Color, etc.**	**Mineral**
5–6	2 (intersecting at right angles)	green to black, Sp. Gr. = 3.2–3.4	Pyroxene (calcium ferromag. silicate)
5.5	2 (intersecting at 60° and 120°)	dark green to black, Sp. Gr. = 3.3	Hornblende (calcium ferromag. silicate)
2.5–3	1 (basal)	dark brown to black Sp. Gr. = 2.8–3.0	Biotite (ferromag. potassium hydrous aluminum silicate)
2	1 (basal)	dark green, Sp. Gr. = 2.6–3.0	Chlorite (ferromag. aluminum silicate)

Nonmetallic Luster, Dark Color (without cleavage)		
Hardness	**Color, etc.**	**Mineral**
7	dark gray to black, conchoidal fracture, Sp. Gr. = 2.65	Flint (silicon dioxide)
7	red, conchoidal fracture, Sp. Gr. = 2.65	Jasper (silicon dioxide)
7	olive green, sugary texture, Sp. Gr. = 3.2–3.5	Olivine (ferromag. silicate)
7	dark red, translucent, Sp. Gr. = 4	Garnet (complex silicate)
2–5	variable dark green and light green, Sp. Gr. = 2.2	Serpentine (hydrous magnesium silicate)

Nonmetallic Luster, Light Color (with cleavage)

Hardness	Cleavage	Color, etc.	Mineral
6	1 (basal)	white to pink to salmon, Sp. Gr. = 2.5–2.6 Streak= white, luster= vitreous	Pink Microcline (potassium aluminum silicate)
4	4 (octahedral)	yellow to colorless to blue to green, Sp. Gr. = 3.2	Fluorite (calcium fluoride)
3	3 (rhombohedral)	white to yellow to colorless to gray; effervesces, Sp. Gr. = 2.7	Calcite (calcium carbonate)
2.5	1 (basal)	colorless to yellow to light brown, Sp. Gr. = 2.8–3.0	Muscovite mica (potassium hydrous aluminum silicate)
2.5	3 (cubic)	colorless to yellow to white to gray, Sp. Gr. = 2.16	Halite (sodium chloride)

Nonmetallic Luster, Light Color (without cleavage)

Hardness	Color, etc.	Mineral
7	colorless to white to gray, glassy luster, conchoidal fracture, Sp. Gr. = 2.65	Quartz (silicon dioxide)
7	light gray, opaque, Sp. Gr. = 2.65	Chert (silicon dioxide)
5	white to yellow to brown, six-sided crystals, Sp. Gr. = 2.3–2.4	Apatite (calcium fluorophosphate)
2	colorless to white, luster is vitreous to silky, pearly, or waxy; white streak Sp. Gr. = 2.31–2.33	Gypsum (calcium sulfate)
1	white to gray to yellow to light green, soapy feel, Sp. Gr. = 2.7–2.8	Talc (hydrous magnesium silicate)